Intelligent Analytics for Industry 4.0 Applications

The advancements in intelligent decision-making techniques have elevated the efficiency of manufacturing industries and led to the start of the Industry 4.0 era. Industry 4.0 is revolutionizing the way companies manufacture, improve, and distribute their products. Manufacturers are integrating new technologies, including the Internet of Things (IoT), cloud computing and analytics, and artificial intelligence and machine learning, into their production facilities throughout their operations. In the past few years, intelligent analytics has emerged as a solution that examines both historical and real-time data to uncover performance insights. Because the amount of data that needs analysis is growing daily, advanced technologies are necessary to collect, arrange, and analyze incoming data. This approach enables businesses to detect valuable connections and trends and make decisions that boost overall performance. In Industry 4.0, intelligent analytics has a broader scope in terms of descriptive, predictive, and prescriptive subdomains. To this end, the book will aim to review and highlight the challenges faced by intelligent analytics in Industry 4.0 and present the recent developments done to address those challenges.

Intelligent Analytics for Industry 4.0 Applications

Edited By
Avinash Chandra Pandey, Abhishek Verma,
Vijaypal Singh Rathor, Munesh Singh, and
Ashutosh Kumar Singh

CRC Press
Taylor & Francis Group
Boca Raton London New York

CRC Press is an imprint of the
Taylor & Francis Group, an **informa** business

First edition published 2023
by CRC Press
6000 Broken Sound Parkway NW, Suite 300, Boca Raton, FL 33487-2742

and by CRC Press
4 Park Square, Milton Park, Abingdon, Oxon, OX14 4RN

© 2023 selection and editorial matter, Avinash Chandra Pandey, Abhishek Verma, Vijaypal Singh Rathor, Munesh Singh, and Ashutosh Kumar Singh; individual chapters, the contributors

CRC Press is an imprint of Taylor & Francis Group, LLC

Library of Congress Cataloging-in-Publication Data
Names: Pandey, Avinash Chandra, editor. | Verma, Abhishek (Professor of
computer science), editor. | Rathor, Vijaypal Singh, editor. |
Singh, Munesh, editor. | Singh, Ashutosh Kumar, editor.
Title: Intelligent analytics for industry 4.0 applications / edited by:
Avinash Chandra Pandey, Abhishek Verma, Vijaypal Singh Rathor, Munesh
Singh and Ashutosh Kumar Singh.
Description: First edition. | Boca Raton : CRC Press, 2023. |
Includes bibliographical references and index.
Identifiers: LCCN 2022055814 (print) | LCCN 2022055815 (ebook) |
ISBN 9781032342412 (hardback) | ISBN 9781032342429 (paperback) |
ISBN 9781003321149 (ebook)
Subjects: LCSH: Industry 4.0. | Decision making—Data processing.
Classification: LCC T59.6 .I56 2023 (print) | LCC T59.6 (ebook) |
DDC 658.4/03028563—dc23/eng/20230302
LC record available at https://lccn.loc.gov/2022055814
LC ebook record available at https://lccn.loc.gov/2022055815

ISBN: 978-1-032-34241-2 (hbk)
ISBN: 978-1-032-34242-9 (pbk)
ISBN: 978-1-003-32114-9 (ebk)

DOI: 10.1201/9781003321149

Typeset in Times
by codeMantra

Contents

Acknowledgments

We express our heartfelt gratitude to CRC Press (Taylor & Francis Group) and the editorial team for their guidance and support during completion of this book. We are sincerely grateful to reviewers for their suggestions and illuminating views for each book chapter presented here in *Intelligent Analytics for Industry 4.0 Applications*.

Editors

Dr. Avinash Chandra Pandey (Member, IEEE) is currently an Assistant Professor in the Discipline of Computer Science & Engineering at PDPM Indian Institute of Information Technology Design and Manufacturing, Jabalpur. He has more than 8 years of teaching and research experience. He has guided many M.Tech. dissertations and B.Tech. projects. He has published more than 20 journal and conference papers in the area of text mining, NLP, and soft computing. His research areas include data analytics, NLP, social network analysis, IoT, cyber-physical systems, and soft computing.

Dr. Abhishek Verma is an Assistant Professor in the Department of Computer Science & Engineering at IIITDM Jabalpur, India (an institution of national importance). He obtained a Ph.D. degree (2020) in the Internet of Things security from the National Institute of Technology Kurukshetra, Haryana, India. He has more than 7 years of experience in research and teaching. He has published several research articles in international SCI/SCIE/Scopus journals and conferences of high repute. He is an editorial board member of Research Reports on Computer Science (RRCS) and an active review board member of various reputed journals, including IEEE, Springer, Wiley, and Elsevier. His current areas of interest include information security, intrusion detection, and the Internet of Things.

Dr. Vijaypal Singh Rathor received his M.Tech. from Maulana Azad National Institute of Technology, Bhopal, and Ph.D. from ABV-Indian Institute of Information Technology and Management Gwalior, India in 2014 and 2020, respectively. At present, he is working as an Assistant Professor in CSE Discipline at PDPM Indian Institute of Information Technology, Design and Manufacturing (IITDM), Jabalpur, India. Prior to joining IIITDM, he worked as an Assistant Professor at Thapar Institute of Engineering and Technology, Patiala, India. He has also worked as an Assistant Professor in CSE Department at Bennett University, Greater Noida, India. Dr. Rathor has also received a grant as a Principal Investigator for the Project entitled "HT-Pred: A complete defensive machine learning tool for Hardware Trojan Detection" from the Data Security Council of India (DSCI). His research interests include hardware security, trustworthy circuit design, hardware Trojan, machine learning, IoT, and cloud computing.

Dr. Munesh Singh received a Ph.D. degree in Computer Science and Engineering from the National Institute of Technology (NIT), Rourkela, India, in 2017. He has industrial experience of one year with the Honeywell Technology Solution Lab, Bangalore, India. He also has academic experience with VIT University from 2017 to 2018 and with IIITDM Kancheepuram from 2018 to 2021. He is currently an Assistant Professor in PDPM Indian Institute of Information Technology, Design and Manufacturing (IIITDM), Jabalpur, India. He has published 14 research articles in IEEE, Springer, and Elsevier journals. He works as a reviewer for Wiley, IEEE, and Springer. His research interests include wireless sensor networks, cloud computing, software-defined networks, industrial IoT, embedded system, and robotics. He is a professional member of IAENG and IEEE.

Dr. Ashutosh Kumar Singh is a Professor and Head in the Department of Computer Applications, National Institute of Technology Kurukshetra, India. He has research and teaching experience in various universities of India, the UK, and Malaysia. He received his Ph.D. in Electronics Engineering from the Indian Institute of Technology, BHU, India and Post Doc from the Department of Computer Science, University of Bristol, UK. He is also charted engineer from the UK. His research area includes verification, synthesis, design and testing of digital circuits, data science, cloud computing, machine learning, security, and big data. He has published more than 330 research papers in different journals, conferences, and news magazines.

Contributors

K. Nageswara Rao Achary
Department of Electronics and
  Communication Engineering
NIST
Odisha, India

Priya Agarwal
Amdocs Development Center India LLP
Haryana, India

Lukman Adewale Ajao
Department of Computer Engineering
Federal University of Technology
  Minna
Minna, Nigeria

P. Saleem Akram
Department of ECE
Koneru Lakshmaiah Education
  Foundation
Guntur, Andhra Pradesh, India

Simon T. Apeh
Department of Computer Engineering
University of Benin
Benin City, Nigeria

Bibhorr
IUBH International University
Bad Honnef, Germany

Subhransu Kumar Das
Department of Electronics and
  Communication Engineering
NIST
Odisha, India

Rajesh Kumar Dash
Department of Electronics and
  Communication Engineering
NIST
Odisha, India

Swastid Dash
NIST
Odisha, India

Ishaan Deep
Department of Computer Science and
  Engineering
Siksha 'O' Anusandhan University
Odisha, India

A. Deiva Ganesh
Indian Institute of Information
  Technology
Kancheepuram, India

Aseem Deshmukh
MIT Pune
Pune, India

S. Sharmila Devi
Department of Electronics and
  Communication Engineering
Sri Ramakrishna Engineering College
Coimbatore, India

Mohit Dua
Department of Computer Engineering
National Institute of Technology
Kurukshetra, Haryana

Shelza Dua
Department of Electronics and
  Communication Engineering
National Institute of Technology
Kurukshetra, Haryana

Piyush Goyal
Discipline of Computer Science &
  Engineering
PDPM IIITDM
Madhya Pradesh, India

Ishu Gupta
National Sun Yat-Sen University
Kaohsiung, Taiwan

Y. Adline Jancy
Department of Electronics and
    Communication Engineering
Sri Ramakrishna Engineering College
Coimbatore, India

Prachi Joshi
VIIT Pune
Pune, India

Sanil Joshi
Department of Computer Engineering
National Institute of Technology
Kurukshetra, Haryana

P. Kalpana
Indian Institute of Information
    Technology, Design and
    Manufacturing
Kancheepuram, India

Neha Katiyar
Noida Institute of Engineering &
    Technology
Greater Noida, India

Bhavana Kaushik
University of Petroleum & Energy
    Studies
Uttarakhand, India

Keshav Kaushik
University of Petroleum & Energy
    Studies, Uttarakhand, India

Animesh Kumar
Discipline of Computer Science &
    Engineering
PDPM IIITDM
Madhya Pradesh, India

M. Lakshmana Kumar
Department of ECE
Koneru Lakshmaiah Education
    Foundation
Guntur, Andhra Pradesh, India

Anjali Kumari
Discipline of Computer Science &
    Engineering
PDPM IIITDM
Madhya Pradesh, India

Priti Kumari
Noida Institute of Engineering &
    Technology
Greater Noida, India

Sanjana Mahapatra
Department of Computer Science and
    Engineering
NIST
Odisha, India

Shrey Maheshwari
MIT Pune
Pune, India

Neha Mandora
MIT Pune
Pune, India

B. A. Manjunatha
Nitte Meenakshi Institute of Technology
Bengaluru, India

Nicole Mathias
Georgetown University
Washington, DC

Shweta Mhaisalkar
MIT Pune
Pune, India

Brojo Kishore Mishra
GIET University
Odisha, India

Sloni Mittal
Hewlett-Packard (HP)
Haryana, India

Rashmy Moray
Symbiosis Institute of Management
  Studies (SIMS)
Symbiosis International (Deemed)
  University
Pune, India

Subhadarshini Mohanty
Department of Computer Science &
  Engineering
Odisha University of Technology AND
  Research
Bhubaneswar, India

Subasish Mohapatra
Department of Computer Science &
  Engineering
Odisha University of Technology AND
  Research
Bhubaneswar, India

Manjushree Nayak
Department of Computer Science and
  Engineering
NIST
Odisha, India

Deepika Padhy
Department of Electronics and
  Communication Engineering
NIST
Odisha, India

Sanjaya Kumar Panda
NIT Warangal
Hanamkonda, Telangana, India

Milind Pande
MIT WPU Pune
India

Avinash Chandra Pandey
Discipline of Computer Science &
  Engineering
PDPM IIITDM
Madhya Pradesh, India

Yadunath Pathak
Visvesvaraya National Institute of
  Technology
Nagpur, India

Amar Patnaik
Symbiosis Institute of Management
  Studies (SIMS)
Symbiosis International (Deemed)
  University
Pune, India

S. Gopal Krishna Patro
GIET University
Gunupur, India
and
Department of CSE
Koneru Lakshmaiah Education
  Foundation
Vaddeswaram, Andhra Pradesh, India

Praveen Pawar
Visvesvaraya National Institute of
  Technology
Nagpur, India

Vishal Pawar
MIT WPU Pune
India

Shyam Sundar Pradhan
Department of Computer Science and
  Engineering
NIST
Odisha, India

Priyanka P. Pratihari
Department of Computer Science and
   Engineering
NIST
Odisha, India

R. Ramya
Department of Electronics and
   Communication Engineering
Sri Ramakrishna Engineering College
Coimbatore, India

Amlan Sahoo
Department of Computer Science &
   Engineering
Odisha University of Technology AND
   Research
Bhubaneswar, India

Mohit Sajwan
Bennett University Greater Noida
Greater Noida, India

Surabhi Sakhshi
Indian Institute of Management
Rohtak, India

Sanskar
Department of Computer Science and
   Engineering
Siksha 'O' Anusandhan University
Odisha, India

Biswa Ranjan Senapati
Department of Computer Science and
   Engineering
Siksha 'O' Anusandhan University
Odisha, India

K. Aditya Shastry
Nitte Meenakshi Institute of Technology
Bengaluru, India

Ashutosh Kumar Singh
National Institute of Technology
Kurukshetra, India

Munesh Singh
Department of Computer Science &
   Engineering
PDPM IIITDM
Jabalpur, India

Payaswini Singh
Department of Computer Science and
   Engineering
Siksha 'O' Anusandhan University
Odisha, India

Simranjit Singh
Bennett University Greater Noida
Greater Noida, India

Jyoti Srivastava
Madan Mohan Malviya University of
   Technology
Gorakhpur, India

Muchharla Suresh
Department of Electronics and
   Communication Engineering
NIST
Odisha, India

Rakesh Ranjan Swain
Department of Computer Science and
   Engineering
Siksha 'O' Anusandhan University
Odisha, India

Ankit Tiwari
TechMatrix IT Consulting Pvt. Ltd.
Uttar Pradesh, India

Ujjwal Tripathi
Discipline of Computer Science &
   Engineering
PDPM IIITDM
Madhya Pradesh, India

Aditi Trishna
Department of Computer Science and
   Engineering
Siksha 'O' Anusandhan University
Odisha, India

Kanishka Vijayvargiya
Discipline of Computer Science &
   Engineering
PDPM IIITDM
Madhya Pradesh, India

Varad Vishwarupe
Amazon, University of Oxford, MIT
   Pune
Pune, Maharashtra

1 Analytics Approach for Intelligent Cyber-Physical System Integration in Industrial Internet of Things (Industry 4.0)

Lukman Adewale Ajao
Federal University of Technology
University of Benin

Simon T. Apeh
University of Benin

CONTENTS

DOI: 10.1201/9781003321149-1

1.1   INTRODUCTION

Internet of Things (IoT) is a worldwide network facility that is capable of connecting billions of devices remotely for interaction and information exchanges over Internet protocol version 6 (IPv6) for low wireless personal area network (6LoWPAN) [1]. The IoT architecture is flourished with resource-constrained devices that may include embedded sensors, wireless connectivity, communication protocols, and information technology (IT) [2]. The communication of these devices depends on the restraint protocols used for the data packet forwarding and routing information through the 6LoWPAN adaptation layer [3]. These protocols include IPv6, 6LoWPAN, and routing protocol for low power and lossy network (RPL). The 6LoWPAN adaptation layer is utilized to overcome the challenges of end-to-end connectivity, and interoperability between both physical and virtual things identities through IEEE 802.15.4 in the network [4]. IoT is a promising emergent technology that provides conveniences, comforts, and quality of life for domestic users, industries, research institutes, and many others through the procuring of smart technology devices in Industry 4.0 [5]. It also helps to create more opportunities in the physical world, better economic practices, and computational ecosystems. While the interaction of things with a human is possible through the IoT communication layers which can be classified into physical/data link, network, application layer, and cloud computing layer [6].

The physical and data link layer utilized Bluetooth low energy (BLE), ZigBee (IEEE 802.11.15), Wi-Fi (IEEE 802.11b/n/g), and many others for communication and routing of big data. The IPv6, 6LoWPAN, RPL, CARPL, or CORPL was adopted over the network adaptation layer for seamless end-to-end connection, communication between the physical environment and logical layers [7]. The routing protocol for low power and lossy network (RPL) was adopted at the lower link network layer, route optimization, and energy management. Thus, the channel-aware protocol (CARP) was used for underwater acoustic communication or in a hazardous environment. The cognitive routing protocol for low power and lossy network (CORPL) was used for cognitive radio or terrestrial communication [8]. However, the application layer functions on the existing constrained application protocols (CoAP) and representational state transfer (REST) for the application program interface (API). This cap is popularly used to conform with constrained devices and for the seamless interaction between the application layer and cloud computing in the IoT networks [9].

The term smart technology in the IoT refers to the physical and logical connectivity adopted for the system conditioning to provide some functions such as remote monitoring, surveillance, and tracking using various sensing technologies (sensors), actuators, robotics, and artificial intelligence [10,11]. The roles of the smart system in the emergent IIoT are varied like passenger trackers, car geolocation monitoring using GPS [12], door access control [13], and security measures using RFID cards [14]. So, smart technology is for green information that connects different industrial technology (IT) products with things related over the network through wireless

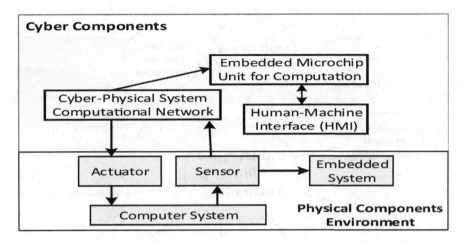

FIGURE 1.1   The cyber-physical system architecture.

devices [15]. Some of these industrial products are cyber-physical systems (CPSs), sensors, artificial intelligence (AI), robotics, and so on.

1.1.1   CYBER-PHYSICAL SYSTEM (CPS)

The CPS is incorporation of different facilities such as digital computing, network-ing, physical objects, and human interfacing systems. This CPS can widely be used for the interrelation of physical components through the human–machine inter-face (HMI) and computational networks to achieve real-time monitoring, big data acquisition, automation, surveillance, and better ecosystem linkages [16]. The CPS implementation comprises sensors, actuators, embedded wireless facilities, and intelligence microchips to achieve various levels of remote-control decisions and far distance monitoring. The illustration in Figure 1.1 shows the CPS design architecture. This efficient design of a CPS has been considered for its suitability in the present IoT and the future emergent IIoT that will accommodate Industry 4.0 technologies.

1.1.2   INDUSTRIAL INTERNET OF THINGS (IIoT)/INDUSTRY 4.0

The IIoT is an enhanced global communication technology that was built to augment the functionalities of IoT operation through the interconnection of sophisticated sen-sors, robotics, CPS, industrial applications, business strategies, and manufacturing products [17, 18]. This is a technology that built a strong merger for adopting AI, cloud computing, real-time data processing, smart factory, and resource planning enterprises over the constraint networks [19]. While Industry 4.0 is the recent fourth industrial revolution technology, its emphases are on the integration of industry man-agement, manufacturing companies, and their products in the IIoT networks [20, 21]. The Industry 4.0 revolution platform accommodates the integration of business processes into the model for resource-rich utilization, technological savvy stakehold-ers, and supply chain management [22, 23]. Others include digitizing information,

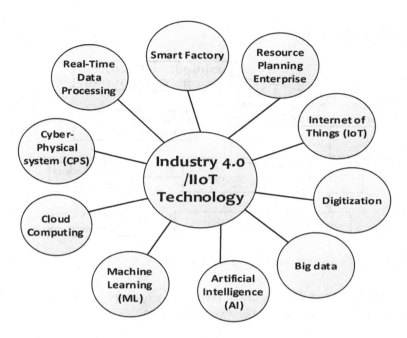

FIGURE 1.2   The components merger in Industry 4.0.

automation, big data, CPS, machine learning for enterprise resource planning, and timely analytics in customer satisfaction. The illustration of components merger in Industry 4.0 is presented in Figure 1.2.

The aim of this emerging fourth industrial revolution (Industry 4.0) was not only to focus on the interconnectivity of smart systems but also to emphasize holistically manufacturing approaches that will provide linkages between physical and digital technology. It will also provide better collaboration between the business vendors, partners, products, and people belonging to Industry 4.0. Meanwhile, interoperability among the smart products is crucial to boost instant data productivity and enhance processes for fast growth.

1.1.2.1   Advantages of Industry 4.0 Technology

The technology of Industry 4.0 is compared with the IIoT, and it delivers a wide range of benefits through the life cycle of CPS products and supply chain management. It includes sharing of real-time information, views of relevant business processes, productions through engineering and design, sales and inventory, field services, and customers [24, 25]. Some other benefits are presented in Figure 1.3.

1.1.2.2   Challenges of Industry 4.0

The incorporation of emergent smart manufacturing or CPS in the IIoT faces some challenges. Industry 4.0 technology focuses are toward the interconnectivity and interoperability of smart IoT products with resource enterprises. The focus also includes better manufacturing companies or ecosystems that will enhance

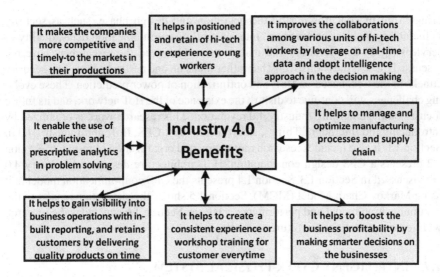

FIGURE 1.3   Illustration of Industry 4.0 benefits.

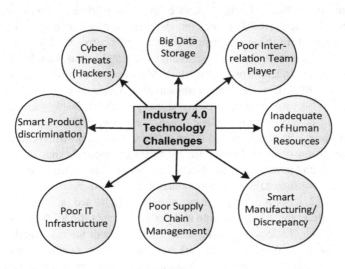

FIGURE 1.4   Industry 4.0 realization challenges.

automation, and real-time big data acquired through the use of machine learning, robotics, and AI with a vision of business strategy that involved people, partners, and products [26]. Figure 1.4 illustrates some of the Industry 4.0 challenges over the emergent IIoT networks.

1.1.3   RESEARCH MOTIVATION

The IIoT platform is an explosive technology that enhances the evolution and availability of smart systems in recent times. But the cooperation of this CPS component brings a lot of concerns during the operation and design phase, such as security vulnerability, power consumption, interoperability, communication delay, quality of services (QoS) deterioration, and many others. However, the researchers have not discussed these issues holistically but rather focus on one pertinent issue like security attacks in the CPS based on IIoT, and optimization of power prediction. These evolving challenges will continue to disturb the existence of the IIoT network, and its future focuses on the emerging Industry 4.0 revolution. This research work is organized by introducing the Internet of Things, smart technology, CPS, IIoT, and Industry 4.0 in Section 1.1. It also includes research motivation and organization of research. Section 1.2 presents a CPS design consideration. IIoT architecture design and Industry 4.0 are discussed in Section 1.3. Section 1.4 presents the CPS communication model in it using Markov Chain Model (MCM). Section 1.5 shows the simulation results of the communication model and performance analysis. Section 1.6 concludes the research with recommendations for future research.

1.2 INTELLIGENT CYBER-PHYSICAL SYSTEM DESIGN CONSIDERATION

The design consideration for this smart embedded wireless system or intelligent cyber-physical system (ICPS) in the fourth generation of an industrial revolution depends on hardware design, software coding, and control theory with a holistic approach of computation to suit physical environmental constraints. For instance, a typical mote in a wireless sensor network (WSN) consists of sensor nodes, a microcontroller unit (MCU), and a wireless transmission device such as a radio frequency (RF) circuit. This mote is mostly powered using a lithium battery or other means of energy and is very hard to replace. Then, the power consumed during sensing and packet transmission is a critical stage in that CPS design, which can shorten the longevity of a sensor node. Others are hardware components that are involved in the designed architecture. Particularly, the Internet of Things (IoT) gateway called a radio or router consumes the maximum amount of power during packet transmission and receiving [27]. So, it is very significant to be enlightened on these analytic factors that can affect the efficient design and performance of the ICPS in the emergent Industry 4.0. The challenges of ICPS design include fault tolerance and reliability, scalability, hardware constraint, network topology design, operating environment, transmission media, routing algorithm techniques, energy consumption, and security.

1.2.1 FAULT TOLERANCE AND RELIABILITY

Fault tolerance is the ability to sustain system functionalities without any interruption, or failure that occurs in any part of the system. This failure occurs as a result of environmental interferences, physical damages, or energy source depletion, which in norms and circumstances should not affect the network. Reliability $R(t)$ can be described as the probability of a system that will continue to survive until it becomes unreliable (failure) $F(t)$ as expressed in equation (1.1). The unreliability of the system and the probability that an embedded system will not fail within the interval $(0, t)$ are expressed as in equations (1.2) and (1.3):

$$R(t) = P(X > t) = 1 - F(t) \qquad (1.1)$$

$$F(t) = 1 - R(t) \qquad (1.2)$$

$$R_j(t) = e^{-\lambda_{jt}} \qquad (1.3)$$

where X is the random variation of the component lifetime, P is the probability, λj is the failure rate of node j, and t is the interval period (time).

1.2.2   SCALABILITY μ (S)

Scalability is the ability to extend or increase the size of a network node without collapsing the network structure and services. It is a process of adding more sensor nodes to a network and still maintaining its efficiency even when several numbers of nodes were added to it. Therefore, a scalable network is a network model that grows with increasing in-network loads as many of these Industry 4.0 products are welcome and integrated into the platform, such as robotic systems, machine learning, AI, big data, techno-preneureship, and many others. The scalability of a network system can be calculated as expressed in equation (1.4):

$$\mu(S) = M\pi T^2 / E \qquad (1.4)$$

where M is the number of scattered sensor nodes in the network environment (E), T is a transmission radio range, and E is the network environment.

1.2.3   HARDWARE SYSTEM DESIGN

Hardware system design refers to the parallel integration or the interconnection of the components, where each component performs its specific function as proved in the analytical models of designs. The hardware component implementation is configured with coding techniques (computational models), which leads to constraints in processor speeds, memory capacity for data storage, power utilization, and hardware failure rates. The ICPS design in the revolution of it includes hardware component integration as a full-functional device (FFD) such as microchips, memory, processors, and power as depicted in the block diagram of Figure 1.5. This system design may also include additional units like a global positioning system (GPS) for location detection and industrial applications.

1.2.4   NETWORK TOPOLOGY

Network topology is a model or structural linkage designed for the controllability of both internal and external sensor nodes, which are interconnected to capture or share empirical properties in a network. Therefore, these topological properties in relating information and profile control between source and sink nodes are constraints when integrated within an embedded wireless system. This constraint is a result of the complexity of the system (density of nodes) and interaction between pairs of nodes

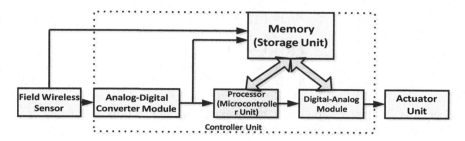

FIGURE 1.5   Block diagram of wireless-based hardware system design.

connected (unweighted edge components) over the networks. Therefore, a suitable topology is important to be adopted for efficient communication in the network due to the various number of sensor nodes involved in the network for surveillance, tracking, and monitoring. The most common network topologies used are star, tree, mesh, and hybrid. Each of these topologies has challenges and better performances in the network as analyzed in Table 1.1, based on their communication range, power consumption, and synchronization period.

1.2.5   Data Model Design

The data model refers to the application requirements for the control of low latency, reading, and writing data with a predefined pointer into a data block [28]. This block of data model required different data types with the predefined schema to store data or a list of values sorted in a table. It includes primitives, collections (lists, maps, and sets), primary key, keyspace (a collection of related tables), and so on. As a result of this illustration in Figure 1.6, efficient data model design for CPS is a critical aspect of smart technology design for better functions in the Industry 4.0 revolution.

1.2.6   Operating Environment

The wireless sensor node is expected to be effectively designed when they are densely deployed directly in contact with the phenomenon or closer to the object for efficient operations and adaptation to the environmental condition. As in robotic system design for a specific application using AI techniques or machine learning for the

TABLE 1.1
Network Topology Analysis

Topology	Range	Power	Period
Mesh	Long	Higher	Not at all
Star	Short	Low	Not at all
Tree	Long	Low	Yes
Hybrid	Long	Very low	Depending on the sensor pattern

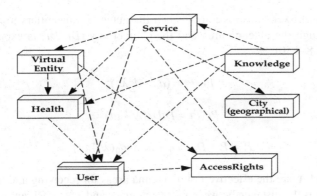

FIGURE 1.6 IIoT data model design networks.

big data acquisition and analysis. In addition, other applications such as underwater acoustic monitoring, surveillance on battlefields, harsh conditions like the nozzle of an aircraft engine, and so on also require effective design of wireless sensor node. The formation of wireless connections and network services in erratic environmental conditions, such as atmosphere/weather, rain, and icing conditions, requires improvement to meet the standard of Industry 4.0 products.

1.2.7 COMMUNICATION AND TRANSMISSION TECHNOLOGY

Signal transmission is very important and crucial to ICPS communication by adopting standardized industrial wireless technologies such as RF and industrial scientific and medical (ISM) bands. This will enhance the efficient communication and interoperability among several things connected over the network of Industry 4.0. In wireless communication systems, the frequency used for transmission affects the amount of data and the speed at which the data can be transmitted. The strength or power level of the transmission signal determines the distance over which the data can be sent and received without errors or loss of signal. In general, the principle that governs this wireless transmission dictates the lower channel frequency can carry less data more slowly over a long distance. The distance between the centers of the cell that uses the same cluster frequency shows the cluster size C and cluster cell. The frequency reuse distance (f_{rd}) is used to normalize the size of each cell in hexagonal shape as expressed in equations (1.5) and (1.6):

$$f_{rd} = \mathrm{sqrt}\{3C\} \tag{1.5}$$

$$C = i^2 + ij^2 + j^2 \tag{1.6}$$

where $C = 1, 3, 4, 7, 9, \ldots$ as imaginable cluster size or single-cell size, and i and j are integers that determine a relative location of coexisting channel cells. Therefore, the equation can be rewritten as follows:

$$f_{rd} = \mathrm{sqrt}\{3(i^2 + ij^2 + j^2)\} \tag{1.7}$$

The effective power transmission between the antenna transmitters to the receiver (P_r, P_t) through the antenna area and their directivity (Dt, Dr) is expressed as in equations (1.8–1.10):

$$Pr/Pt = (a_r a_t / d^2 \lambda^2)^2 \tag{1.8}$$

$$Pr/Pt = DtDr(\lambda/4\pi d)^2 \tag{1.9}$$

$$P_r = P_t + D_t + D_r + 20\log_{10}(\lambda/4\pi d)^2 \tag{1.10}$$

where A_r and A_t are the effective radio antenna areas for receiving and transmitting the signal, d is the distance between radio antennas, and λ is an RF antenna.

1.2.8 ENERGY CONSUMPTION

The battery life of ICPS can be significantly extended if sensor nodes are configured to coordinate or dictate when the entire circuit should be ON or OFF or by using a meta-heuristic algorithm. A sensor node in the ICPS can be configured to determine when the entire circuit should begin monitoring an event, by activating the circuit or else should be put into deep sleep mode for energy savings. The lifetime of WSN is highly dependent on the power available at each node in the network called active power (Active$_p$). Without providing any reverse energy, this power is draining off during operation. It means energy constraint in an embedded wireless system [29, 30]. The average power (Av_p) consumption of a sensor node and the sensor node duty cycle (D_{node}), which is the fraction of time (t) when the node is active and the period is T, can be expressed as in equations (1.11–1.13):

$$Av_P = \text{Active}_P * D_{node} \tag{1.11}$$

$$D_N = \text{Active}_t / T \tag{1.12}$$

$$T = \text{Active}_t + \text{Sleep}_t \tag{1.13}$$

However, one of the most challenges of this ICPS component is a power constraint because each node or component in the architecture will be loaded and expended energy of the information flow. The energy expended by each node of ICPS in the network will be dependent on the functions (loads) which can be affected by location (topology) and material properties (operating characteristics). The power consumption of the node and the entire network can be estimated. Let us consider a single ICPS $\Psi_{ps} = (\phi_n, \xi_s)$ with components (nodes) integrated and determine the load ζ, but explicit service classifications are disregarded, and faulty physical node (component) is not considered. The total load of the ICPS per unit time is given as equation (1.14), and the structural load of ICPS node j up-to-the-minute time step n is expressed as in equation (1.15). However, the constraint of this system occurs when there is an excess flow of information transmission that causes a node buffer overflow and packet loss [31] as given in equation (1.16):

$$\zeta \Psi_{ps} = \sum_{j \in \wedge_{p-normal}} \alpha.K_{nj}^{\delta} \tag{1.14}$$

$$\zeta_{nj}^{c} = \frac{\zeta \Psi_{ps}.d_{nj}^{\delta}}{\sum_{j \wedge_{p-normal}} d_{nj}^{\delta}} \tag{1.15}$$

$$\zeta_{nj}^{c} \leq (1 + \sigma_{c}).\zeta_{0j}^{c} \tag{1.16}$$

where Ψ_{ps} is the CPS, ϕ_{n} is predictable of nodes, ξ_{s} is the established node edge, $\Lambda_{p-normal}$ is the standard process in the physical network environment, κ^{δ} is the sensor node degree, and j is the number of nodes in the network up-to-the-minute time step n, while κ and δ are arbitrary variables in the control of the CPS. σ_{c} is the node tolerance coefficient and ζ^{c} is the load that is added to the sensor node j for the system faulty identification.

1.2.9   ROUTING PROTOCOLS

A wireless sensor packet routing is relying on the network layers of the IIoT stack model, while a routing algorithm is adopted to discover the best routes for packet transmission and to ensure the safe delivery of data packets to the destinations in the networks. The transport control protocol/Internet protocol (TCP/IP) is mostly responsible to guarantee packet delivery and establishing a reliable communication connection between two hosts in the network. It also repairs and maintains the routes disruption when radio links (or hops) along established routes are broken. This problem can ensue due to sensor node relocation or node failure, server RF interference, and congestion.

Also, the routing processes in the network are very challenging due to several characteristics that distinguish them from contemporary communication and wireless ad-hoc networks as in Table 1.2. These include the building of global addressing and routing algorithms for classical IP-based protocols. The application of sensor nodes in the network is to sense data from multiple regions (sources) and route to a particular sink or gateway (destination), which consumes energy and bandwidth utilization. Therefore, the nodes that are tightly deployed in a constrained network environment will have limited or constrained resources like power consumption, processing capacity, and data storage [32]. The HEED and LEACH energy-aware routing protocols that function as a multihop routing network algorithm using transmission power of intercluster communication to select cluster head (CH) among the sensor nodes in a network are expressed as in equation (1.17) [33]:

$$n_{\text{prob}} = C_{\text{prob}} * \varepsilon_{r}^{i} / \varepsilon_{T} \tag{1.17}$$

where $\hbar_{\text{prob}}$ is the probability of selecting a sensor node as a CH, C_{prob} is the maximum number of clusters probability in the network, ε_{r}^{i} is the residual energy of an ith node, and ε_{T} is the total energy of the network. However, routing is a critical concern in the distributed sensor network when considering the homogeneity in the way packets are routed between source and destination across varying network topologies.

TABLE 1.2
Routing Algorithm and Its Functions

Protocols	BSN	Mobility	Functions
LEACH	1	Stable BS	The distributed cluster node is formed to extend the node's lifetime.
EWC	1	Stable BS	Distributed cluster is formed to guarantee data delivery and life improvement.
SPIN	1	Movable	Exchange metadata to reduce the number of messages and lifetime.
REAR	>1	Restricted	Lifetime extended and data delivery insured.
Energy aware	>1	Restricted	Lifetime extended.
Direct diffusion	>1	Restricted	It establishes efficient n-way communication paths for fault tolerance.
Information driven	1	Restricted	It saves more energy using optimization of direct diffusion techniques.
Gradient	1	Restricted	It delivers data through the minimal number of hops.
PEGASIS	1	Stable base station	It is lifetime with bandwidth optimization.
Energy-aware based cluster head	1	Not at all	It is lifetime and operates in a real time.
Self-organized	1	Movable	It improves fault tolerance.
Minimum energy communication	1	Not at all	It is lifetime and self-reconfiguration.
Geographic adaptive fidelity	>1	Restricted	It increases the network lifetime as the number of nodes increases.

But, the security mechanism for this activity (routing process) is reflected as more problematic and interesting compared to others.

1.2.10 SECURITY CHALLENGES

The significant function of a secure mechanism in the IIoT network requires the urgent attention of the experts due to billions of things and smart manufacturing components that are expected to be flourishing in the universal network shortly. This IIoT provides many conveniences and benefits to humans in the wide areas of smart technologies. But the security challenge and resource constraint devices will make it unsafe, and it may become an untrustworthy network [34]. The integration of ICPS into the IIoT model is susceptible to different attacks due to weak security counter-measures at each layer of the network, and it has become a serious challenge to the existence of networks. Any provision to counter these security challenges with a robust cryptographic method or advanced encryption algorithm may result in net-work overhead. Other challenges include energy inefficiency, latency (delay), and packet loss with throughput degradation because of their computation power, which is not suitable for such constraint devices. So, any nodes that are compromised by

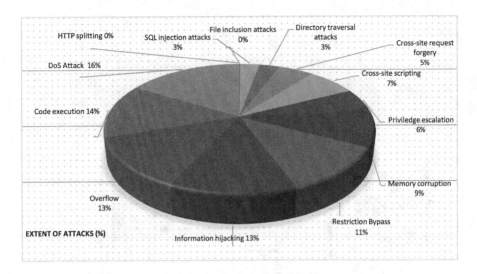

FIGURE 1.7   Analysis of securities attack measurement in the IIoT environment.

malicious actors, hackers, or social networks intruder can affect the performance of the ICPSs in the network. The consequence of these malicious nodes in the network is the misleading routing of sensor node traffic, generating false alarm reports, message modification, false route identification, masquerading, and breaching the trust, privacy, and confidentiality of the network information [35].

This significant security flaw in IIoT networks threatens the existence of ICPS functionality, and it requires the urgent attention of experts due to billions of things and machines that are expected to be flourishing in the Industry 4.0 network shortly. Also, it is noted that smart devices are increasing in the market with several threats that are replicated to destroy the existence of industrial revolution worldwide networks. These include bonnets, data spoofing, eavesdropping, ransomware, social engineering, denial of service (DoS), and distributed denial of service (DDoS). The extent of the cyber threat effect on smart technology devices is analyzed as shown in Figure 1.7. This IIoT network threat is established in a distributed network during the routing protocol initialization through an identified route from one ICPS node to another until the destination is discovered.

1.3   THE INDUSTRIAL INTERNET OF THINGS ARCHITECTURE DESIGN AND INDUSTRY 4.0

The IIoT design and Industry 4.0 technology include the integration of different facilities and services (such as embedded system (ES), network technology (NT), and IT) for sensing, and control of physical objects by linking them to cyberspace network [36]. Thus, these three technologies are very crucial to the functionality of Industry 4.0 architecture. Other expected technologies are robotics, AI, machine learning, big data, digital computing, smart manufacturing, supply chain, business strategies, and

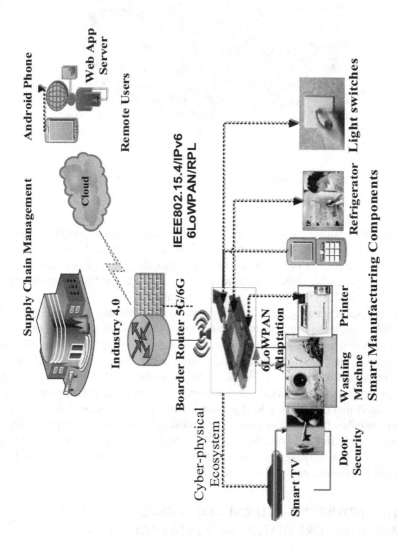

FIGURE 1.8   Intelligent cyber-physical system implementation in IIoT.

industrial products [37, 38]. The implementation of Industry 4.0 architecture in the IIoT network illustrated in Figure 1.8 is based on the following technologies:

i. Embedded system can be a subset of IIoTs that provide service networks such as a software package and firmware for industrial products. This industrial product may include automobiles, electronics, industrial control systems, biometric systems, medicare devices, digital signal devices, and robotics. The embedded system architecture may include a microchip, gateway, sensor/mote, GSM module, and others to render functions like the QoS, end-to-end communication, and interoperability.
ii. Network technology in this context supports end-to-end connectivity in the IIoT network using wired technology or wireless protocols. The role of this technology allows a seamless distribution of resources and application programs over the Internet. The network service is used for the building of infrastructure that will relay information to the cloud database through an information system, Internet, voice calls, and GSM/SMS.
iii. IT is a subsection of information and communication technology that supported the computing application, processing, storage, retrieving, and handling of information over cyberspace. The IIoT information service includes mobile web applications, programming, and a cloud database system.

1.4   THE COMMUNICATION MODEL OF ICPS IN INDUSTRY 4.0

The communication model of ICPS depends on the protocol handler which utilized TCP/IP for end-to-end communication and encapsulation of data from the local area network to the remote interfaces. The communication protocol of low-level firmware implementation is targeted to function over a wide area of network connectivity due to the gateway connection between the cloud database and remote server using a wireless access point (WAP) and wireless gateway support node (WGSN). In the process, the dynamic session renegotiations (DSRs), forced session renegotiations (FSRs) on the Internet gateway, and content management system server (CMSS) are to significantly improve the sensor nodes' connectivity over the Internet and reduce packet loss rates among the sensor nodes.

The DSR functions in two ways: exchange packet transmission and verifying the status of both gateway and CMS server channels (either in uplink or downlink status). The DSR also verifies the appearance of socket failure or deadlock of a packet sent between the clients and servers. If the data packets are unable to deliver or received from a TCP/IP connection, the FSR procedure begins to check available packets or data received from the gateway unit for a period of elapsed time. This process will cause CMSS to wait for a new renegotiation or reconnection between the client, server, and closed TCP/IP socket. But, if network service or packet is lost during the transmission and communication setup between gateway and CMSS, the TCP/IP will attempt a reconnection until connection establishment is successful. The involvement of these different application layer protocols and programming language design is used for communication between IoTs gateway and cloud database

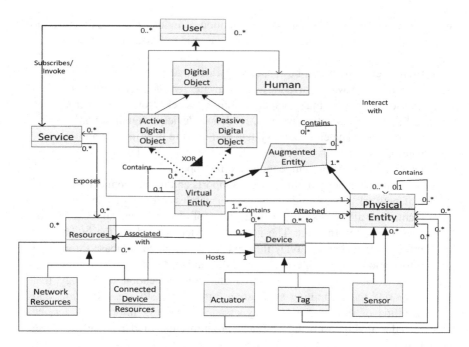

FIGURE 1.9 The UML illustration of the cyber-physical system communication model in Industry 4.0.

using java native language (JNL), Micro-C, MySQL, and many protocols like CoAPs, MQQT, and HTTP. The UML of the CPS communication model in Industry 4.0 is given in Figure 1.9.

1.5　SIMULATION RESULTS OF CYBER-PHYSICAL COMMUNICATION MODEL IN IIOT ENVIRONMENT

The communication model of the CPS in the IoT architecture was developed and simulated in Cooja-Contiki software. The CoAP web service with RESTful HTTPS was adopted as an interoperable application-level protocol. These protocols focus on end-to-end communication, and interoperability achievement among things (nodes) connected in a network. Cooja-Contiki is an Internet of Things–supported simulation tool supported with C-language for application program development through a java application interface (JNI). It supports both large and small networks of sensor motes to be simulated and implemented with objective functions and system settings like generation of packets rate, media access control (MAC) protocols, throughput, packet loss, and network topology. This CPS network design consists of four sensor motes that run CoAP servers as resource-constrained protocols, and they are sporadically queried by the network client using CoAP protocol. The GETs request command is sent to the network server in the simulation setting to acquire the parameter from the border router and help to connect the network with the Internet using the Contiki-tunslip utility. The communication of these wireless sensor nodes simulation results shows the accuracy and efficiency of the packet transmitted from the field (physical layer) to the cloud database system. The results of packet generation (packet/sec) and packet loss against throughput during packet transmission are presented in Tables 1.3 and 1.4, respectively. Figures 1.10 and 1.11 depict graphs of

TABLE 1.3
Simulation Results of Packet Rate Generation Against Throughput

	Network Throughput (kb/s)			
	Sensor	Sensor	Sensor	Sensor
PGR (packet/s)	Node1	Node2	Node3	Node4
1	0.915	0.8757	0.7254	0.6505
5	1.2714	0.9932	0.7792	0.6566
10	1.9454	1.3552	1.1732	1.8463
15	2.122	1.4545	1.1921	1.2571
20	2.2748	1.9837	1.3534	1.5462

TABLE 1.4
Simulation Results of Packet Rate Generation Against Packet Loss

	Packet Loss (%)			
	Sensor	Sensor	Sensor	Sensor
PGR (packet/s)	Node1	Node2	Node3	Node4
1	1.20	2.70	4.55	5.50
5	3.90	5.71	8.55	11.61
10	5.25	7.02	12.25	21.77
15	6.42	7.05	13.45	14.55
20	7.51	9.21	14.6	17.05

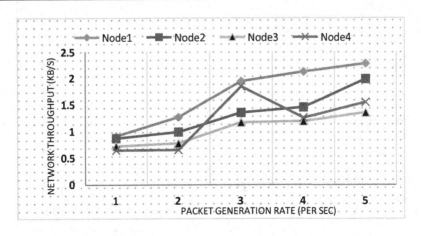

FIGURE 1.10 Network throughput against packet generates rate.

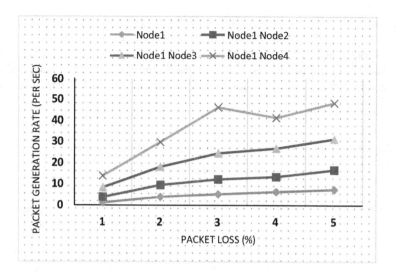

FIGURE 1.11   Packet generates rate against packet loss.

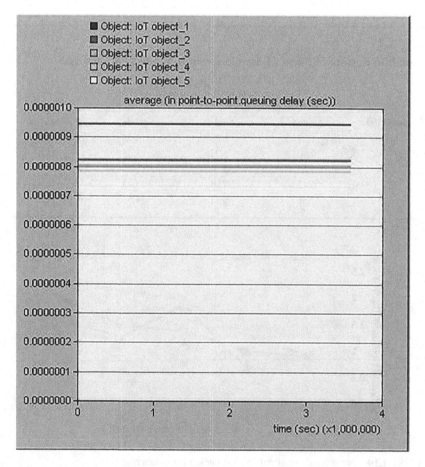

FIGURE 1.12   Average packet delay in point to point.

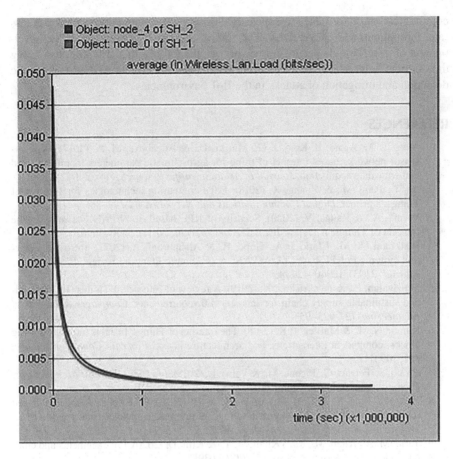

FIGURE 1.13   Average period/packet transmits (bits/s).

packet generation and packet loss against throughput, respectively. Figures 1.12 and 1.13 show the average packet delay and the average period of packet transmission during end-to-end communication nodes in the simulation environment.

1.6 CONCLUSION

This research analyzed the selected operational factors that affect the design of the CPS and constraints of the IIoT model. The research survey exposed that energy consumption, fault tolerance, reliability, and scalability affect the optimal performance of the CPS-based IIoT. It also gives a clear and cohesive description of the proposed design model together with the pertinent systemic components of the CPS. The design requirements and technical procedures for integrating embedded devices, sensors, actuators, and wireless connectivity for the enhancement of the IoT architectural model are also elucidated in this work. The efficient design of a CPS depends on hardware design, software coding, and control theory with the holistic approach of computation to suit physical environmental constraints. The simulation of ICPS communication performance was measured using packet generation rate, packet loss rate, throughput, and packet delay. The future works will be focused on the development of a lightweight secure framework for the CPS in Industry 4.0. Also, a big data storage vulnerability to cyber attacks demanded an optimized approach for the detection and mitigation of attacks in the IIoT environments.

REFERENCES

1. Ajao, L. A., Agajo, J., Kolo, J. G., Maliki, D., & Adegboye, M. A. (2017). Wireless sensor networks based-internet of thing for agro-climatic parameters monitoring and real-time data acquisition. *Journal of Asian Scientific Research*, 7(6), 240–252.
2. Ai, Y., Peng, M., & Zhang, K. (2018). Edge computing technologies for Internet of Things: a primer. *Digital Communications and Networks*, 4(2), 77–86.
3. Verma, A., & Ranga, V. (2020). Security of RPL based 6LoWPAN Networks in the Internet of Things: A review. *IEEE Sensors Journal*, 20(11), 5666–5690.
4. Pasikhani, A. M., Clark, J. A., Gope, P., & Alshahrani, A. (2021). Intrusion detection systems in RPL-based 6LoWPAN: A systematic literature review. *IEEE Sensors Journal*, 21(11), 12940–12968.
5. Manavalan, E., & Jayakrishna, K. (2019). A review of Internet of Things (IoT) embedded sustainable supply chain for industry 4.0 requirements. *Computers & Industrial Engineering*, 127, 925–953.
6. Kumar, N. M., & Mallick, P. K. (2018). The Internet of Things: Insights into the building blocks, component interactions, and architecture layers. *Procedia Computer Science*, 132, 109–117.
7. Cirani, S., Ferrari, G., Picone, M., & Veltri, L. (2018). *Internet of Things: Architectures, Protocols and Standards*. John Wiley & Sons: Hoboken, NJ.
8. Yang, Z., Ping, S., Sun, H., & Aghvami, A. H. (2016). CRB-RPL: A receiver-based routing protocol for communications in cognitive radio enabled smart grid. *IEEE Transactions on Vehicular Technology*, 66(7), 5985–5994.
9. Tariq, M. A., Khan, M., Raza Khan, M. T., & Kim, D. (2020). Enhancements and challenges in coap—A survey. *Sensors*, 20(21), 6391.
10. Darby, S. J. (2018). Smart technology in the home: time for more clarity. *Building Research & Information*, 46(1), 140–147.
11. Brougham, D., & Haar, J. (2018). Smart technology, artificial intelligence, robotics, and algorithms (STARA): Employees' perceptions of our future workplace. *Journal of Management & Organization*, 24(2), 239–257.
12. Aliyu, S., Yusuf, A., Abdullahi, U., & Hafiz, M. (2017). Development of a low-cost GSM-bluetooth home automation system. *International Journal of Artificial Intelligent and Application*, 8(8), 41–50.
13. Jimoh, O. D., Ajao, L. A., Adeleke, O. O., & Kolo, S. S. (2020). A vehicle tracking system using greedy forwarding algorithms for public transportation in urban arterial. *IEEE Access*, 8, 191706–191725.
14. Cheng, J., Chen, W., Tao, F., & Lin, C. L. (2018). Industrial IoT in 5G environment towards smart manufacturing. *Journal of Industrial Information Integration*, 10, 10–19.
15. Goddard, N. D. R., Kemp, R. M. J., & Lane, R. (1997). An overview of smart technology. *Packaging Technology and Science: An International Journal*, 10(3), 129–143.
16. Humayed, A., Lin, J., Li, F., & Luo, B. (2017). Cyber-physical systems security–A survey. *IEEE Internet of Things Journal*, 4(6), 1802–1831.
17. Boyes, H., Hallaq, B., Cunningham, J., & Watson, T. (2018). The industrial internet of things (IIoT): An analysis framework. *Computers in Industry*, 101, 1–12.

18. Malik, P. K., Sharma, R., Singh, R., Gehlot, A., Satapathy, S. C., Alnumay, W. S., ... & Nayak, J. (2021). Industrial Internet of Things and its applications in industry 4.0: State of the art. *Computer Communications*, *166*, 125–139.
19. Bai, C., Dallasega, P., Orzes, G., & Sarkis, J. (2020). Industry 4.0 technologies assessment: A sustainability perspective. *International Journal of Production Economics*, *229*, 107776.
20. Serpanos, D., & Wolf, M. (2018). Industrial Internet of Things. In *Internet-of-Things (IoT) Systems* (pp. 37–54). Springer: Cham.
21. Ghobakhloo, M. (2020). Industry 4.0, digitization, and opportunities for sustainability. *Journal of Cleaner Production*, *252*, 119869.
22. Dalenogare, L. S., Benitez, G. B., Ayala, N. F., & Frank, A. G. (2018). The expected contribution of Industry 4.0 technologies for industrial performance. *International Journal of Production Economics*, *204*, 383–394.
23. Vaidya, S., Ambad, P., & Bhosle, S. (2018). Industry 4.0–A glimpse. *Procedia Manufacturing*, *20*, 233–238.
24. Xu, L. D., & Duan, L. (2019). Big data for cyber physical systems in industry 4.0: A survey. *Enterprise Information Systems*, *13*(2), 148–169.
25. Enrique, D. V., Druczkoski, J. C. M., Lima, T. M., & Charrua-Santos, F. (2021). Advantages and difficulties of implementing Industry 4.0 technologies for labor flexibility. *Procedia Computer Science*, *181*, 347–352.
26. Castelo-Branco, I., Cruz-Jesus, F., & Oliveira, T. (2019). Assessing Industry 4.0 readiness in manufacturing: Evidence for the European Union. *Computers in Industry*, *107*, 22–32.
27. Lozano, C. V., & Vijayan, K. K. (2020). Literature review on cyber physical systems design. *Procedia Manufacturing*, *45*, 295–300.
28. Stavroulaki, V., Kelaidonis, D., Petsas, K., Moustakos, A., Vlacheas, P., Demestichas, P., & Hashimoto, K. (2016). D2. 1 Foundations of Semantic Data Models and Tools, IoT and Big Data Integration in Multi-Cloud Environments, University of Surrey, 1–115.
29. Ajao, L. A. Agajo, J., Mua'zu, M. B., & Schueller, J. (2020). A scheduling-based algorithm for low energy consumption in smart agriculture precision monitoring system. *Agricultural Engineering International: CIGR Journal*, *22*(3), 103–117.
30. Ajao, L. A., Agajo, J., Gana, K. J., Ogbole, I. C., & Ataimo, E. E. (2017). Development of a low power consumption smart embedded wireless sensor network for the ubiquitous environmental monitoring using ZigBee module. *ATBU Journal of Science, Technology and Education*, *5*(1), 94–108.
31. Qu, Z., Xie, Q., Liu, Y., Li, Y., Wang, L., Xu, P., Zhou, Y., Sun, J., Xu, K., & Cui, M. (2020). Power cyber-physical system risk area prediction using dependent Markov chain and improved grey wolf optimization. *IEEE Access*, *8*, 82844–82854.
32. Manzoor, B., Javaid, N., Rehman, O., Akbar, M., Nadeem, Q., Iqbal, A., & Ishfaq, M. (2013). Q- LEACH: A new routing protocol for WSNs. *Procedia Computer Science*, *19*, 926–931.
33. Touati, Y., Ali-Chérif, A., & Daachi, B. (2017). Routing Information for Energy Management in WSNs. In Touati, Y., Ali-Cherif, A., & Daachi, B. (Eds), *Energy Management in Wireless Sensor Networks* (pp. 23–51). Elsevier: Amsterdam, The Netherlands.
34. Nigam, G. K., & Dabas, C. (2021). ESO-LEACH: PSO based energy efficient clustering in LEACH. *Journal of King Saud University-Computer and Information Sciences*, *33*(8), 947–954.
35. Yu, X., & Guo, H. (2019, August). A survey on IIoT security. In *2019 IEEE VTS Asia Pacific Wireless Communications Symposium (APWCS)* (pp. 1–5). IEEE.
36. Jhanjhi, N. Z., Humayun, M., & Almuayqil, S. N. (2021). Cyber security and privacy issues in industrial Internet of Things. *Computer Systems Science and Engineering*, *37*(3), 361–380.

37. Ngabo, D., Wang, D., Iwendi, C., Anajemba, J. H., Ajao, L. A., & Biamba, C. (2021). Blockchain-based security mechanism for the medical data at fog computing architecture of internet of things. *Electronics*, *10*(17), 2110.
38. Frank, A. G., Dalenogare, L. S., & Ayala, N. F. (2019). Industry 4.0 technologies: Implementation patterns in manufacturing companies. *International Journal of Production Economics*, *210*, 15–26.

2 Digital Twins A State of the Art from Industry 4.0 Perspective

Ganesh A. Deiva and P. Kalpana
Indian Institute of Information Technology

CONTENTS

2.1 INTRODUCTION

In the present data-centric environment, organizations are focusing on the transformation toward digitalization. It is leading to numerous modifications in production processes, enhancing communication and cooperation. This kind of transformation for creating responsive, flexible, and visible networks is known as digital transformation. This is the basis of the present industrial revolution that brings a paradigm shift to manufacturing. For instance, Industry 4.0 builds on the foundation of the Internet of Things (IoT) to enable monitoring the real-time systems, develop cyber-physical frameworks, and offer the integration between the real and virtual world [1]. It is more about intelligent, real-time communication, and self-adapting that integrates all areas of manufacturing systems such as engineering, manufacturing, logistics, and sales [2]. It aims to facilitate more efficient, higher quality, and faster production. Given this, digital twin (DT) is a virtual model of a real-time system based on a simulation, which optimizes the system for enhanced efficiency. This DT theory has evolved from a conceptual model for product life cycle management to an important tool for decision-making in this digital environment [3]. These models offer opportunities for researchers and industry experts to achieve smart, sustainable

DOI: 10.1201/9781003321149-2

manufacturing systems, efficient supply chain management (SCM), and intelligent mechanisms to predict, assess, and manage sudden disruptions. The global pandemic COVID-19 has driven many organizations toward automated production processes using IoT, faster decision-making against disruption, and accurate prediction in production planning. The virtual design and optimization of intelligent systems are the significant pillars for the successful transformation to Industry 4.0. Some of the significant inabilities of the conventional industry practices are as follows [4]:

- The optimization of the product design outcome (gap between design and manufacturing).
- The improvement of the production processes concerning the dynamic environment.
- The prediction of unprecedented events across the various echelons of the supply chain (SC).
- Real-time monitoring of the quality, performance, transportation, and the information sharing.

These observations highlight the necessity to transform the internal structure in the organizations to improve product quality, enhance process monitoring, and mitigate uncertainty. In industries, DTs can be introduced at any stage in the production process. They are capable of predicting and optimizing the product and product lines in addition to the preproduction planning and design. Nowadays, they are exploited in the modeling of the advanced SC that can assess strategies supporting SC optimization. Recent studies highlighted the significance of developing digital SC models that represent the actual SC digitally. The potential of DTs and the recently emerging technologies demand further investigations to understand the challenges, applications, and under-explored avenues in Industry 4.0.

The residual part of this chapter provides the structure as follows: we present the bibliometric analysis of the extant literature in the next section. Then, we illustrate the significant components of DT architecture. Some notable applications are discussed in Section 2.4. The last section concludes the study.

2.2 RECENT STUDIES ON DIGITAL TWIN

This section illustrates the bibliometric analysis of recent studies on DT technologies. Bibliometric analysis is a publication statistical research method. Researchers have exploited this method frequently to evaluate scientific publications systematically and identify the research gaps in a specific area.

The authors selected 'Science direct' as the data source. 'Digital twin' and 'Industry 4.0' were chosen as the keywords. We restricted the articles that are published only in the year 2022 to capture the recent advancements concerning Industry 4.0. A total of 395 articles have been identified and exploited for this analysis. The journals like *Computers in Industry, Computers and Industrial Engineering, Journal of Manufacturing Systems, Advanced Engineering Informatics*, and *International Journal of Production Economics* are some of the significant contributors. The results were reexplored using the R tool and Biblioshiny software.

Figure 2.1 illustrates the treemap chart according to the keywords of the recent articles. This chart contains the frequency of the essential terms and keywords that have been mentioned in the selected articles, such as Industry 4.0, DT, smart manufacturing, AI, sustainability, digitalization, predictive maintenance, IoT, blockchain, Machine Learning (ML), simulation, and augmented reality. All major significant terms related to DT components and applications have been covered in the selected articles. The co-occurrence of the keywords network is shown in Figure 2.2. It can be understood from the diagram that the majority of the DT and Industry 4.0 applications have been linked with 'smart manufacturing', 'sustainability', 'predictive maintenance', and 'circular economy'. Based on the selected articles, clusters form in the network of keywords. It indicates that the study has focused on investigating DT and Industry 4.0. Similarly, the major components involved in DT technologies such as IoT, AI, ML, blockchain, cyber-physical systems (CPSs), simulation, and big data analytics have been highlighted in the network diagram.

Among the collection, the articles that are discussing concepts like digitalization, sustainability, DT, and Industry 4.0 have been published more in number. The size of the circle represents the density of the publications with respect to the key terms (Figure 2.3).

Word TreeMap

industry 4 0	digital twin	smart manufacturing	simulation	deep learning	additive manufacturing	digital transformation	manufacturing	
		artificial intelligence	sustainability	digital twins	augmented reality	cyber physical systems	control	
					blockchain	discrete event simulation	iot	
	machine learning	digitalization	internet of things	predictive maintenance	circular economy	human robot collaboration	systematic literature review	
							virtual reality	

FIGURE 2.1   Keywords treemap.

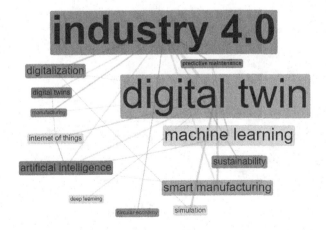

FIGURE 2.2   Co-occurrence network.

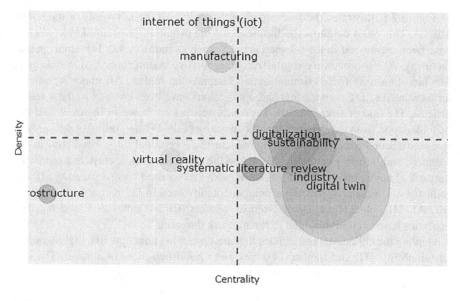

FIGURE 2.3 Keywords density map.

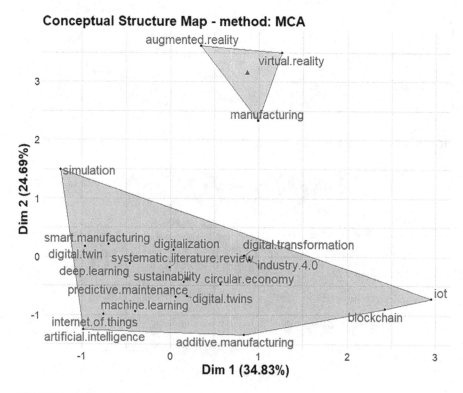

FIGURE 2.4 Conceptual structure map.

The key terms that are associated with the selected studies are categorized into several clusters to comprehend the significant avenues that influence DT in Industry 4.0. The conceptual structure map shown in Figure 2.4 is developed by multiple correspondence analysis (MCA) using Biblioshiny software. The chart illustrates the distribution of the key terms according to their position on the map. The terms with the same dimension seem nearer on the map. Based on these, the authors consolidate the studies into two notable clusters such as DT applications and visualization technologies.

As a result of the bibliometric analysis, the components and applications involved in DT architecture are discussed in the following sections:

2.3 COMPONENTS OF DIGITAL TWIN

This section presents the notable Industry 4.0 technology components in the development of DT architecture (Figure 2.5).

- **IoT:** The IoT for interconnected systems drives the significant interest in adopting DT frameworks in organizations. Conventional manufacturing systems are incapable of getting/accessing crucial data and dealing with the large volume of data. Presently, the advancements in IoT allow industries to access, organize, and assess complex unstructured data technically and economically. It drives DT in numerous industries by providing mass production efficiency, mass customization, and mass personalization [5]. The extant studies defined the industrial Internet of things (IIoT) as a smart infrastructure that integrates physical and virtual environments to furnish advanced operations. For example, Zhao et al. developed a DT-based architecture for tracking real-time information using IoT [6]. Most manufacturing industries have automated their processes using these IoT-based

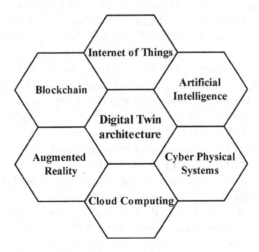

FIGURE 2.5 The components of digital twin framework.

advancements, generating digital information. Recently, IoTwins have been explored to implement cost-effective DTs with the support of cloud, ML, edge, and IoT in Industry 4.0 settings [7].

- **Artificial intelligence (AI):** Advanced data analytics, one of the prominent AI techniques, has enhanced the quality of information extraction from unstructured data [8]. Various AI/ML algorithms support DTs to identify future uncertain events and make faster decision-making. Integrating AI with manufacturing technology helps industry practitioners to develop data-driven decision-making systems. Cavalcante et al. investigated the incorporation of ML and simulation to develop DTs to aid in the development of mitigation strategies related to supply disruption [9]. Today's DT models are primarily confined to understanding and rudimentary reasoning of decisions for designing the smart manufacturing system. Integrating the advanced AI algorithms into the DT models in the future could result in achieving the potential to conceive and develop new design solutions automatically. Given this, AI is integrated with IoT and CPS for illustrating the virtual representation of DT in real time [10].

- **Cloud computing:** The productivity and efficiency can be enhanced by enclosing manufacturing resources in an open environment like cloud computing. It is applicable in various domains for developing business architectures, and it is popular among researchers and industry experts. The exploitation of DTs is a data-intensive process, which involves high computing and storage requirements. Therefore, the necessity of advanced systems for real-time information monitoring and the integration of cloud into conventional manufacturing systems are significant challenges in implementing DT systems [11]. Advanced data analytics plays a crucial role in assessing the transfer of data to cyberspace.

- **CPS:** CPS is using the IoT to integrate disguised knowledge of production systems. These systems have the potential to predict operational decisions in smart manufacturing without the need for physical setup. However, the important challenges in implementing CPS are real-time synchronization and dynamic decision-making. Numerous CPS-based frameworks have been addressed in the literature. Web-based frameworks for increasing productivity, distributed production frameworks to adapt to the dynamic environment, and stochastic production systems to attain real-time control in making decisions are some of the significant applications. Therefore, the existing studies highlight that DTs are the technical fundamental for developing CPS in Industry 4.0 [12].

- **Blockchain:** DT is a centralized architecture that lacks traceability, audit, and trusted data province. Blockchain offers decentralized and collaborative manufacturing processes. Therefore, incorporating IIoT with blockchain can support a decentralized manufacturing network. The studies addressed that the physical verification and mapping of manufacturer relationships can be overcome by exploiting blockchain and smart contracts [13]. Also, they have been adopted to secure data sharing in fog-based IoT systems [14]. These advancements are expected to track, and monitor the transactions, improve data security, and share efficiency in DTs. In Industry 4.0, numerous research has been done on blockchain-enabled smart manufacturing applications. However, practical concerns like scalability and flexibility of systems have hampered the adoption of blockchain in Industry 4.0, and need to be addressed further [15].

- **Augmented reality:** The impact of DTs is quite powerful when the outputs are delivered in 3D. This requires the transformation of 2D systems like computers and screens to high-end 3D display technologies such as augmented reality. Virtual reality in the recent DT frameworks provides productive integration between the available digital information and the physical system. However, the lack of knowledge about power and training, and complex system incorporation are some of the notable barriers to proposing augmented reality-based DT frameworks [16]. Scholars are attempting to enhance design efficiency by incorporating DT systems with augmented reality/virtual reality technologies.

2.4 APPLICATIONS OF DIGITAL TWIN

The above discussion reveals the essential components of a DT framework. This section discusses the notable applications of DT technologies.

2.4.1 SMART MANUFACTURING

In this data-centric world, the manufacturing process is transforming toward data-centric and perception-enabled smart manufacturing. This involves the creation and exploitation of data that addresses the integration of information technology and data analytics. DT helps to achieve on-demand manufacturing services such as smart factory layout, scheduling, smart manufacturing, and human–robot collaboration by acting as an enabler or direct assistance [17]. The literature describes how a DT can be exploited to achieve the collaboration of humans and robots at a higher level [18]. Because of DTs, it is now possible to analyze human thinking and behavior in real time using theoretical aspects [19]. Zhuang et al. presented a four-layer smart manufacturing architecture for complex assembly lines [20]. It includes managing the synchronization of real-time data, building an assembly floor DT, big data-enabled predictive analysis, and exploiting DTs to assess services. It has been discovered that adopting a DT in production processes can help save energy and resource [21]. Despite its impressive advances, the current DT model is still inadequate to satisfy the demands of the accurate interconnection of the virtual model/data and physical systems [22]. The development of feasible standards and proven architecture is a critical step toward the success of DT-smart manufacturing.

2.4.2 SUPPLY CHAIN MANAGEMENT

In SCM, new paradigms, concepts, and frameworks are being accelerated by digital technologies. The growth of digital SCs and intelligent process management is aided by IoT, CPSs, and smart, connected devices. The introduction of DTs for automating enterprises in the direction of SC integration is essential; however, it is lacking. In addition, the lack of real-time insights and flexible designing/planning systems makes DT adaptation problematic. According to recent studies designing and developing a digital SC twin that digitally replicates the organization's actual SC is essential [23]. To achieve advanced SC modeling based on prescriptive analytics that operates in real time and can decide the proper strategies supporting SC optimization, a digital SC twin is exploited. Since it relies on AI and simulation approaches proposed by individual developers (vendors), it is impossible to copy or generate them later using different developers. The DT's promise in SCs as well as current advances in new technologies like additive manufacturing, BDA, and IoT, necessitates a systematic review of the extant literature to comprehend DT components, applications, and challenges. As previously discussed, a DT depicts the present state of SCs, by incorporating real-time transportation, inventory, demand, and capacity information. For instance, if a strike occurs at an international logistics hub, a risk monitoring tool can detect the disruption and send it to the simulation model as a disruptive event for further assessment. The simulation part in the DT can aid in predicting and quantifying the risk propagation. Furthermore, it enables rapid testing and adoption of recovery strategies and contingency measures. The studies addressed that the investigations regarding the adoption of DT in production logistics are still in a nascent stage [24].

2.4.3 SUSTAINABILITY

The steady escalation of environmental problems, including carbon emissions and nuclear pollution, has necessitated a transition in industries from traditional broad economic growth to sustainable development in recent years. To assist product design, manufacturing, efficient service, and sustainability enhancement, DTs incorporate physical, virtual, and connected product data. Given this, a simulation-based DT framework contributes to sustainable development and circularity by assessing the resource efficiency and recovery of high-quality secondary resources [25]. In general, the role of DTs is important in delivering a data-rich runtime for achieving sustainability. Also, concerning the evaluation of sustainable ideas, DTs based on data fusion concepts can aid in reducing ambiguity and uncertainty. With the support of ML algorithms, sustainability for a better life cycle can be addressed. For instance, concerning smart city infrastructure, we can explore the noise levels in industries and residential areas [26]. Similarly, the integration of ML and IoT has potential benefits in building environmental monitoring systems [27]. The extant literature highlights that the DTs are becoming the critical component of a product, liabilities, and industrial operations in order to accomplish long-term goals.

2.4.4 INTELLIGENT RISK MANAGEMENT

Predictive maintenance is a preventative strategy for predicting unprecedented events well in advance of incompetence to avoid them. Because of its potential to respond to uncertainty developing in real-time scenarios, DT appears to be more beneficial than conventional approaches that had prior empirical data. Ivanov et al.

proposed a decision-support system that incorporates data analytics, simulation, and optimization [28]. The optimization allows proactive, resilient SC optimization, and the simulation provides the assessment of SC dynamic behavior in the case of anticipated disruptions. Furthermore, this enables reactive, predictive modeling of disruption consequences on SC performance and of recovery strategies, which are then optimized using an analytical model in a prescriptive way. On the other hand, the system's data analytics component is intended to identify disruptions in real time utilizing process feedback data [29]. Given this, Deiva Ganesh and Kalpana developed a conceptual DT-based framework to provide a holistic intelligent risk management mechanism [30]. In general, SCs often are highly coordinated yet they are stressed and lead to failures when an unprecedented event occurs. DTs can help simulate these sudden rises and drops in supply, in order to meet the production demands and make the whole process seamless. To achieve this, Big data analytics (BDA) and AI provide an entirely new potential benefit to data-driven risk management. Due to its dynamic tracking, this technique shows promise in minimizing unscheduled maintenance and lowering excessive expenses. Predictive maintenance using DT extends its application to aerospace industries, and healthcare processes also [31].

2.5 CHALLENGES AND IMPLEMENTATION ISSUES

This section highlights the significant challenges and issues related to DT implementation. However, as the interest in DT research is increasing rapidly, the studies addressed the challenges found in terms of infrastructure, data availability, technological and security issues, expectations, trust, and modeling [32], as follows:

- Similar to AI, blockchain, and IoT, the next primary challenge is infrastructure development. Since data analytics, virtual reality tools, and IoT facilitate the successful framework of a DT, it requires infrastructure that permits the success of the aforementioned technologies. The absence of well-connected smart networks, information sharing, and monitoring devices hampered the implementation of these frameworks in most industries.
- The current level of data standards and structures is the next important challenge. DT requires quality data with constant, and continuous data availability. Therefore, inspecting, handling, and storing such a huge volume of data need efficient data management technologies. Inconsistent and noisy data may lead to the failure or poor performance of DTs in attaining their objectives. Given this, an extensive analysis is needed for the right data collection and its exploitation for achieving an effective outcome.
- In continuation, the data processing requirements will increase due to the exponential growth of information growth. Hence, to provide a greater level of processing speed, the data analytics algorithms like ML and deep learning need high-end processors. The lack of advanced technology adoption can hinder the progress of introducing DTs in organizations.
- Concerning Industry 4.0, privacy and security issues are the important threats with DT adoption. Because of the large volume of data, they are vulnerable to cyber and hacker attack issues. The annual risk report (2021)

reported that the global pandemic increased the trust issues related to cyber risks like data breaches to 73%. Therefore, emerging technologies like data analytics and IoT for DT implementation should follow the standard, security policies, updating practices, and ethical guidelines to overcome the trust issues.

- Further investigation on DT technology is required to overcome the trust issues from both industries and users' perceptions. Verifying and validating the performance is crucial for reducing trust-related issues. Hence, this enabling technology will provide valuable insights with more understanding into the security and ethical-related steps.

- Despite the infrastructure, data processing, and privacy challenges, some notable specific issues relating to the design and modeling of DT frameworks also should be addressed. The current practices are lacking in standardized modeling and domain-specific modeling approaches. Following standardized approaches in modeling can improve user understanding while providing information flow across the different stages and implementation of a DT.

2.6   FUTURE RESEARCH AVENUES

An extensive future investigation is required to ensure the users and organizations understand the ultimate benefits of a DT. In addition, the significant challenges that are addressed need to be considered to attain the full potential of DT in various domains. Some of the promising avenues for further research are as follows:

i. Though the literature addressed integrated and manufacturing DTs, they are not realized in industries. There is a need for research in modeling DT that exploits cyber-physical fusion.

ii. DT technology requires further research in fault diagnosis and predictive maintenance for industrial processes.

iii. DTs for monitoring the health and well-being of humans through simulating the impact of lifestyle changes.

iv. Prognostics and health management is one of the most promising areas with DT through the integration of advanced data analytics.

v. Concerning data collection, processing, and analysis for DTs, methodologies should be explored to enhance accuracy. Data fusion, remote surgery, and modeling for healthcare applications facilitate more opportunities for future research.

vi. DTs for smart city development and traffic management systems are another advanced open research area to be addressed. In such cases, exploring standardized modeling approaches is essential.

vii. In addition to manufacturing excellence, the scope of DTs for proactive decision-making in SCM is increasing significantly to ensure SC operation continuity.

viii. Deriving intelligent and holistic risk management mechanisms that combine data analytics, simulation, and optimization is an emerging area to minimize the impact of deep disruptions like COVID-19.

ix. Developing high-fidelity dynamics models through AI-equipped DT to support complex robotic systems is under-explored in the extant literature.

x. Transitioning to high-end technologies such as augmented reality and virtual reality can be an exciting avenue to enhance the interaction between the physical and virtual environment.

2.7 CONCLUSION

In Industry 4.0, smart manufacturing has become an emerging direction for the global manufacturing industry. This chapter surveys how the DTs are integrated into the applications such as smart manufacturing, SCM, sustainable designs, and intelligent risk management mechanisms. In addition, critical components such as IoT, CPSs, virtual reality, data analytics, AI, blockchains, and cloud computing for supporting the DT are analyzed. With ready access to them, most sectors have begun to exploit DTs to manage their key assets. In addition, industry practitioners can use DTs to anticipate operational issues, enhance product quality, and minimize downtime. Currently, DT applications are primarily directed toward achieving manufacturing excellence. Expanding the reach of DTs to cover all the things and SC practitioners will enable them to make more proactive decisions. With increased awareness of the economic benefits of adopting this technology in the next few years, it is expected to serve as the backbone in the Industry 4.0 era in accelerating industrial transformation. Future work could include an assessment of DT technologies in the sectors of renewable energy, smart cities and mobility, and healthcare.

REFERENCES

1. Kamble, S. S., Gunasekaran, A., Parekh, H., Mani, V., Belhadi, A., & Sharma, R. (2022). Digital twin for sustainable manufacturing supply chains: Current trends, future perspectives, and an implementation framework. *Technological Forecasting and Social Change*, *176*, 121448.

2. Zambrano, V., Mueller-Roemer, J., Sandberg, M., Talasila, P., Zanin, D., Larsen, P. G., … Stork, A. (2022). Industrial digitalization in the industry 4.0 era: Classification, reuse and authoring of digital models on Digital Twin platforms. *Array*, *14*, 100176.

3. Cinar, Z. M., Nuhu, A. A., Zeeshan, Q., & Korhan, O. (2019, September). Digital twins for Industry 4.0: A review. In *Global Joint Conference on Industrial Engineering and Its Application Areas* (pp. 193–203). Springer, Cham.

4. Jiang, Y., Yin, S., Li, K., Luo, H., & Kaynak, O. (2021). Industrial applications of digital twins. *Philosophical Transactions of the Royal Society A*, *379*(2207), 20200360.

5. Stavropoulos, P., & Mourtzis, D. (2022). Digital twins in industry 4.0. In *Design and Operation of Production Networks for Mass Personalization in the Era of Cloud Technology* (pp. 277–316). Elsevier, Amsterdam, The Netherlands.

6. Zhao, Z., Shen, L., Yang, C., Wu, W., Zhang, M., & Huang, G. Q. (2021). IoT and digital twin enabled smart tracking for safety management. *Computers & Operations Research*, *128*, 105183.

7. Costantini, A., Di Modica, G., Ahouangonou, J. C., Duma, D. C., Martelli, B., Galletti, M., … Cesini, D. (2022). IoTwins: Toward implementation of distributed digital twins in Industry 4.0 settings. *Computers*, *11*(5), 67.

8. Li, X., Cao, J., Liu, Z., & Luo, X. (2020). Sustainable business model based on digital twin platform network: The inspiration from haier's case study in China. *Sustainability*, *12*(3), 936.

9. Cavalcante, I. M., Frazzon, E. M., Forcellini, F. A., & Ivanov, D. (2019). A supervised machine learning approach to data-driven simulation of resilient supplier selection in digital manufacturing. *International Journal of Information Management*, *49*, 86–97.

10. Radanliev, P., De Roure, D., Nicolescu, R., Huth, M., & Santos, O. (2022). Digital twins: Artificial intelligence and the IoT cyber-physical systems in industry 4.0. *International Journal of Intelligent Robotics and Applications*, *6*(1), 171–185.

11. Lu, Y., & Xu, X. (2019). Cloud-based manufacturing equipment and big data analytics to enable on-demand manufacturing services. *Robotics and Computer-Integrated Manufacturing*, *57*, 92–102.

12. Liu, C., Jiang, P., & Jiang, W. (2020). Web-based digital twin modeling and remote control of cyber-physical production systems. *Robotics and Computer-Integrated Manufacturing*, *64*, 101956.

13. Reddy, K. R. K., Gunasekaran, A., Kalpana, P., Sreedharan, V. R., & Kumar, S. A. (2021). Developing a blockchain framework for the automotive supply chain: A systematic review. *Computers & Industrial Engineering*, *157*, 107334.

14. Mohapatra, D., Bhoi, S. K., Jena, K. K., Nayak, S. R., & Singh, A. (2022). A blockchain security scheme to support fog-based internet of things. *Microprocessors and Microsystems*, 89, 104455.

15. Leng, J., Wang, D., Shen, W., Li, X., Liu, Q., & Chen, X. (2021). Digital twins-based smart manufacturing system design in Industry 4.0: A review. *Journal of Manufacturing Systems*, *60*, 119–137.

16. Kamble, S. S., Gunasekaran, A., & Gawankar, S. A. (2018). Sustainable Industry 4.0 framework: A systematic literature review identifying the current trends and future perspectives. *Process Safety and Environmental Protection*, *117*, 408–425.

17. Cimino, C., Negri, E., & Fumagalli, L. (2019). Review of digital twin applications in manufacturing. *Computers in Industry*, *113*, 103130.

18. Malik, A. A., Masood, T., & Bilberg, A. (2020). Virtual reality in manufacturing: Immersive and collaborative artificial-reality in design of human-robot workspace. *International Journal of Computer Integrated Manufacturing*, *33*(1), 22–37.

19. Ardanza, A., Moreno, A., Segura, Á., de la Cruz, M., & Aguinaga, D. (2019). Sustainable and flexible industrial human machine interfaces to support adaptable applications in the Industry 4.0 paradigm. *International Journal of Production Research*, *57*(12), 4045–4059.

20. Zhuang, C., Liu, J., & Xiong, H. (2018). Digital twin-based smart production management and control framework for the complex product assembly shop-floor. *The International Journal of Advanced Manufacturing Technology*, *96*(1), 1149–1163.

21. Kannan, K., & Arunachalam, N. (2019). A digital twin for grinding wheel: An information sharing platform for sustainable grinding process. *Journal of Manufacturing Science and Engineering*, *141*(2), 1–14.

22. Cheng, J., Zhang, H., Tao, F., & Juang, C. F. (2020). DT-II: Digital twin enhanced Industrial Internet reference framework towards smart manufacturing. *Robotics and Computer-Integrated Manufacturing*, *62*, 101881.

23. Simchenko, N. A., Tsohla, S. Y., & Chuvatkin, P. P. (2020). Effects of supply chain digital twins in the development of digital industry. *International Journal of Supply Chain Management*, *9*(3), 799–805.

24. Kaiblinger, A., & Woschank, M. (2022). State of the art and future directions of digital twins for production logistics: A systematic literature review. *Applied Sciences*, *12*(2), 669.

25. Bartie, N. J., Cobos-Becerra, Y. L., Fröhling, M., Schlatmann, R., & Reuter, M. A. (2021). The resources, exergetic and environmental footprint of the silicon photovoltaic circular economy: Assessment and opportunities. *Resources, Conservation & Recycling*, *169*, 105516.

26. Bhoi, S. K., Mallick, C., Mohanty, C. R., & Nayak, R. S. (2022). Analysis of noise pollution during Dussehra festival in Bhubaneswar smart city in India: A study using machine intelligence models. *Applied Computational Intelligence and Soft Computing*, *2022*, Article ID 6095265.
27. Bhoi, S. K., Panda, S. K., Jena, K. K., Sahoo, K. S., Jhanjhi, N. Z., Masud, M., & Aljahdali, S. (2022). IoT-EMS: An Internet of Things based environment monitoring system in volunteer computing environment. *Intelligent Automation and Soft Computing*, 32(3), 1493–1507.
28. Ivanov, D., Dolgui, A., Das, A., & Sokolov, B. (2019). Digital supply chain twins: Managing the ripple effect, resilience, and disruption risks by data-driven optimization, simulation, and visibility. In *Handbook of Ripple Effects in the Supply Chain* (pp. 309–332). Springer, Cham.
29. Ivanov, D., & Dolgui, A. (2021). A digital supply chain twin for managing the disruption risks and resilience in the era of Industry 4.0. *Production Planning & Control*, *32*(9), 775.
30. Deiva Ganesh, A., & Kalpana, P. (2022). Future of artificial intelligence and its influence on supply chain risk management–A systematic review. *Computers & Industrial Engineering*, *2022*, 108206.
31. Liu, J., Zhao, P., Zhou, H., Liu, X., & Feng, F. (2019). Digital twin-driven machining process evaluation method. *Computer Integrated Manufacturing Systems*, *25*, 1600–1610.
32. Fuller, A., Fan, Z., Day, C., & Barlow, C. (2020). Digital Twin: Enabling technologies, challenges and open research. *IEEE Access*, *8*, 108952–108971.

3 Human-Centered Approach to Intelligent Analytics in Industry 4.0

Varad Vishwarupe
Amazon, University of Oxford, MIT Pune

Milind Pande and Vishal Pawar
MIT WPU Pune

Prachi Joshi
VIIT Pune

Aseem Deshmukh, Shweta Mhaisalkar,
Shrey Maheshwari, and Neha Mandora
MIT Pune

Nicole Mathias
Georgetown University

CONTENTS

3.1 INTRODUCTION

Transformation of industrial machines, brought about by the collaboration of automation technology, machine learning, and communication abilities for the independent running of an industry, is the concrete objective of the Industry 4.0 framework. Industry 4.0 plays an important role in the manufacturing industry by ensuring security, flexibility, customization, time efficiency, and a dynamic environment as well as increased productivity and quality [1,2]. Through connecting smart devices and machinery, employing self-learning solutions, and enhancing self-direction

DOI: 10.1201/9781003321149-3

capabilities, it is envisioned that the communication cost is reduced while flexibility for manufacturing, mass customization capabilities, production speed, and quality are increased [3–5].

Though the number of benefits of Industry 4.0 is tempting, Industry 4.0 also reveals several challenges or in a more critical way obstacles to improving productivity at the workplace [6]. One such challenge is the work division between manpower and machines, which allows us to explore Industry 4.0 catered human–computer interaction (HCI) to reap the benefits of automated and smart machines, without compromising on the strength of the current manpower. Work division would enable a role change of humans to shift from low-level operations—which can be dangerous, dirty, difficult, and dull tasks—to high expertise and safe tasks [7–10]. Moreover, human intelligence and intervention remain a key role because of the safety, security, social aspects, and uncertainties posed by such autonomous systems [11–14]. Even with the dangerous roles out of the way of industry workers with the fourth industrial revolution, a good design and HCI in place would be required to make humans adept in the newly defined roles, so as to operate complex machines with negligible or no error. Thus, this research work aims to review the current vision of HCI and UX Design of Industry 4.0 and recommends practices best suited for the fourth industrial revolution.

3.2   RELATED WORK

With the increasing stress on human satisfaction, ease and focus on better experiences, there has been an emerging trend in research work related to human-centered design and development of HCI from an Industry 4.0 perspective. Extracts from such work are mentioned in this section, for obtaining a better picture of the overall context. A good user experience (UX) and intuitive HCI is the cornerstone for the smooth operation of any industry, and with Industry 4.0 mechanisms, these become crucial. A system needs to be designed to generate positive UX for increasing user association and encouragement [15–19] presented that bad user interface design can lower user motivation to use a system. Beard-Gunter [20] worked with HCI design in industries to develop and optimize engagement metrics compared to games. To ensure the effectiveness of a system, a proper balance is required to be maintained between usability and system functionality [21]. Considering design elements in developing the industrial automation system can facilitate the meaningful interactions between the system and the user [22,23].

Recent advances in artificial intelligence (AI) and HCI have seen remarkable contributions that have shaped the world. From being a virtual concept in scientific fiction (sci-fi) a few years ago, AI is literally changing the way we act and interpret the world around us. HCI and AI can be coined as two sides of the same coin according to us and can be termed as entailing the same story in different ways or entailing different parts of the same story, but ultimately working in unison for a common goal. Ben Shneiderman, one of the pioneers in the field of HCI, has emphasized the use of AI in collaboration with HCI researchers to further develop the field of human-inspired AI [3,24].

There has also been a huge upheaval in the way HCI is used in the manufacturing industry to optimize for quality and not cost. The work of Uzor et al. [25] is particularly intriguing in this regard wherein they have tweaked a very important facet of Bayesian optimization, the activation function, and a particularly important statistical machine learning method to improve user interface design using crowdsourcing. The use of virtual reality (VR) and mixed reality (MR) as a driving force for the domain of HCI has been on the rise too [5,6]. Using intuitive user interfaces (IUIs), VR and MR gadgets, and head-mounted displays has also made it possible to interact with the physical medium in a way that is more immersive. Simulation of touch displays for people with motor disabilities, gesture typing, 3D-based VR models, and gaze-based motion tracking gadgets are also some of the advancements that have helped in fusing the two realms of AI and HCI, by developing smart and interactive systems [7–11].

The use of AI-based HCI systems and HCI-oriented AI systems has also been observed, especially in the medicine sector. AI-based diagnosis of difficult-to-identify diseases and in areas where human intervention seems to have stuck at a dead end enables clinicians to find out ways which are improving healthcare such as Sio and Hoven [14], Weichhart et al. [15] works on gender classification in task functional brain networks. Use of AI–HCI conjunct systems in improving language models for natural languages also depicts the pervasive nature of the said field. Isen [16] Islam et al. [17] have done some remarkable work in this regard. Inferring web page relevance, using HCI models in certain Internet of Things (IoT) tasks, generating personalized recommendations for certain users in the browser, and also using few facets of HCI in the development of smart set-top box TV recommendation systems have enabled the development of smart AI-based systems [12,18–22,26,27].

While there has been a substantial work at the crossroads of AI and HCI, it is still not extensive enough to be able to use HCI and AI in scenarios wherein it becomes difficult to have experts from both fields working together such as cyber-physical systems [24,28–30]. Thus, it is important that the stakeholders from respective fields are shown what, when, why, and how they can contribute, what the major roles of each discipline and its experts are and how can AI be used as a catalyst for developing HCI systems and vice versa [26,27]. In the context of this paper, we try to gather insights from the previous works in the aforementioned and present our research that shall help develop this exciting field of knowledge.

When it comes to HCI in Industry 4.0, we need to understand the prior work in the context of a plethora of subfields, viz. human–machine interaction (HMI), virtual, augmented, and MR applications in the context of cyber-physical systems, and surveys which have been conducted keeping HCI and HMI under the purview on Industry 4.0 as a research topic [31,32]. A large number of surveys covering peculiar aspects of HMI or human factors in Industry 4.0 has been identified in the endeavor of collecting relevant literature for this paper. The vast majority of these studies are specifically oriented toward either VR or augmented reality (AR) applications, or both, within Industry 4.0 operations and, thus, cover only a subset of this paper's scope [33,34]. Büttner et al. [35] conduct a survey on AR and VR applications in Industry 4.0 manufacturing activities, more precisely on the available platform technologies and application areas, creating a small-scale design space for such MR applications in manufacturing. This design space differentiates among four general application scenarios and four types of MR technology platforms available for application, respectively.

The application scenarios are manufacturing, logistics, maintenance, or training while the available platforms comprise mobile devices (AR), projection (AR), and head-mounted displays (HMDs) (AR or VR). Dini and Dalle Mura [36] present surveys on AR applications, however not restricted to Industry 4.0-related application scenarios. They investigated general commercial AR applications including industrial scenarios also, among others, civil engineering. The specific AR application scenarios which they examine based on related scientific literature are maintenance and repairing, inspection and diagnostics, training, safety, and machine setup. Nahavandi [8], in turn, examine scientific research on AR applications for assembly purposes from a time span of 26 years starting as early as 1990 and concentrating mainly on the period from 2005 until 2015. Thus, they extend the scope to many years before the advent of Industry 4.0-related initiatives and ideas. The major application purposes of AR in assembly tasks that they investigate are assembly guidance, assembly design and planning, and assembly training [37,38].

Choi et al. [39] provide a literature review on AR applications with an even less specific focus on industrial deployment by examining the state of the art at the time in research on collaboration in AR considering a wide range of possible application fields, the industrial sector being only one of those. As a result, they identify remaining research challenges relating to collaboration in AR which are the identification of suitable application scenarios and interaction paradigms as well as an enhancement of the perceived presence and situational awareness of remote users.

Palmarini et al. [40], in turn, conduct a structured literature review on different software development platforms and types of data visualization and hardware available for AR applications in various maintenance scenarios The aim and purpose of their study is to derive a generic guideline facilitating a firm's selection process of the appropriate type and design of AR application, tailored to the specific type of maintenance activity at hand which the firm is planning to enhance utilizing AR technology. Lastly, Choi et al. [39] provide surveys on VR technology in an industrial environment, the latter group of authors concentrating on a potential combination of VR technology with discrete event simulation for scenario testing in Industry 4.0 activities. Hermann et al. [41] on the other hand, present a survey on VR applications in manufacturing, concentrating on potential contributions of VR deployment in the development process for new products and deriving a mapping of different types of VR technology toward the different steps of the product development process. Therefore, Dini and Mura [36] consider the applicability of various VR technologies for the phases of concept development, system-level design, design of details, testing and refinement, and launch of production. Besides those surveys on AR and VR applications, a literature review by Hecklau et al. [42] exists on the major challenges as well as skills and competencies needed for future employees under an Industry 4.0 scenario. The authors utilize the insights from the literature analysis to structure the required skills according to different categories, based on which a competence model is created analyzing employees' levels of skills and competencies, which will be particularly important in an Industry 4.0 working environment. The main categories for the competence model are technical, methodological, social, and personal competencies.

Uzor et al. [25], Palmarini et al. [40], Zahoor et al. [43,44] and Vishwarupe et al. [45,46] also present a very lucid case study of using Amazon Mechanical Turk and crowdsourcing platforms for facilitating HCI studies in the context of Industry 4.0 applications, wherein the importance of having a synergy between the user interface for crowdsourcing based studies and industrial requirements is enunciated. While we have tried to cover the maximum bases for our work related to HCI and HMI in the context of Industry 4.0, it is important to highlight that this list of related work is not all-encompassing. Instead, it is rather supposed to provide an outlook of other existing scientific literature regarding the topic of HCI and Industry 4.0 and thereby reaffirm that, to the best of the author's knowledge, a study matching of this magnitude ceases to exist. We have strived to provide a very detailed bird's eye view of the state of the art in HCI and Industry 4.0 issues, which intersect at the crossroads of both disciplines. This aspect facilitates the justification of this key research and signifies the necessary importance as well as the relevance of this research paper. Though UX and human-centered design have been explored in different contexts, there is a dearth of studies and research on its consideration, use, and impact on an Industry 4.0 setup. This chapter focuses on the exploration of human-centered design principles and customized HCI to tend to the complexities involved in Industry 4.0 and the role of humans in such an ecosystem.

3.3   RESEARCH QUESTIONS

At the outset, we need to base our study on the fundamental understanding of what HCI is and how it is relevant under the purview of Industry 4.0. In its entirety, HCI is the study of the interaction between humans and computers, particularly as it pertains to the design of technology. HCI is at the crossroads of user-centered design, user interface (UI), and UX to create intuitive products and technologies. Researchers who specialize in HCI think about conceptualizing and implementing systems that satisfy human users. HCI also helps to make interfaces that increase productivity, enhance UX, and reduce risks in safety-critical systems. This is especially relevant in heavy industrial applications of Industry 4.0 such as manufacturing and cyber-physical systems. Therefore, HCI is on the rise for developing intelligent interactive systems. While defining the research questions for this paper, it is important to identify the key principles of interaction. The gold standard for this is the Norman's model of interaction. Norman's interaction model is a noteworthy and pioneering model for HCI-based studies. It proposes that a user first establishes a goal and then performs actions using the system to achieve that goal. Thereafter, the system provides the result of the user actions on the UI. A particular user then minutely assesses this result and sees if their objective has been achieved or not. If not, a new goal is established, and the cycle is repeated. This model of interaction explained is divided into eight primary stages (Figure 3.1):

This model helps us to understand where things go haywire in our designs. Under the context of Industry 4.0, there are issues when it comes to the sections of machine

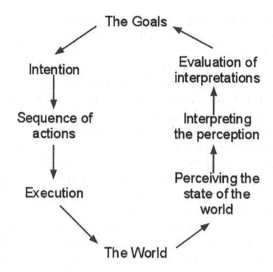

FIGURE 3.1   Norman model of interaction.

execution and problem evaluation by the cyber-physical systems. Machines perceive the world in binaries, as opposed to sensory inputs that are so readily perceived by humans. There are also issues with the interpretation of perception which further causes issues with the evaluation of those interpretations. There is a measurable difference between user actions and those that a particular system can perform. A potent and engaging interface allows a user to perform an action without system limitations. This is especially important for Industry 4.0 which is nothing but an interconnected web of system of subsystems consisting of IoT and cyber-physical systems. The second consideration wherein the importance of having a proper interaction mechanism between humans and machines for cyber-physical systems is the difference between the presentation of an output and the user's expectations. An effective interface can be easily evaluated by a user, but it fails to evaluate when only machines are involved in the process. As a preliminary step, few research questions were chalked out to define the basis of this research.

The focus of these research questions is to primarily study the current visions of an Industry 4.0 ecosystem and especially explore this industrial revolution from the perspective of humans. Answering the formulated research questions would provide an overarching view of the human-centered approach of building the next industrial revolution. For answering the research questions, domain-specific keywords from each research question were used to explore the current research work through Google Scholar. Google Scholar was preferred as the source of all research content, as it gave an expansive search result and encompassed research work published from a wide range of publication organizations, thus covering a broad spectrum of resources. Research material from a diverse set of publication houses—IEEE, Research Gate, Elsevier, Science Direct, and MDPI—were studied and analyzed. Providing better user support whenever there is a tangible difference expected output and the output generated by the user can be a vital factor in deciding the end goal of an efficient cyber-physical system. Thus, we pose the following important three aspects and five questions and try to answer them with our study on the principles of HCI under the context of intelligent analytics for Industry 4.0:

- The UI and UX design paradigm:
  What facets of the UI–UX components of the HCI be envisioned from an Industry 4.0 perspective?
- The interaction paradigm:
  How should the interaction mechanism between humans and machines be shaped under the context of Industry 4.0?
- The manpower–machinery paradigm:
  How will manpower be trained to interact with, supervise, and maintain complex machinery of Industry 4.0?
  1. Why is there a need to involve humans in the process of developing this Industry 4.0–HCI confluence?

     Naturally, it is a human tendency to question things and have a reasoning that is both acceptable as well as viable when it comes to perceiving the world around us. The question of why to involve humans in the Industry 4.0–HCI confluence is vital because autonomous intelligent systems can falter at times, where only human intervention works.
  2. Which stakeholders need to be present in the development of the Industry 4.0–HCI confluence?

     Traditionally, AI has been looked upon as a branch of mathematics and computer science, which tries to mimic the reasoning abilities of the human brain. However, with the pervasive nature of AI and its encroachment on our day-to-day life, it is essential to broaden the scope of involving different stakeholders in this major confluence.
  3. When to trust the AI in Industry 4.0 and when to trust humans, in the decision-making at the Industry 4.0–HCI confluence?

     Gradually, with the advent of large amount of data that is available, issues pertaining to trust, ethics, privacy, and security of user's data have been on the rise. It is important to be able to distinguish when to trust the AI and when not to, and rely on human judgment instead. This research question delves into the ethical side of AI and HCI.
  4. Where can the users work in tandem with the AI/where they cannot, pertaining to the confluence?

     Eventually, the choice of where to use AI-enabled systems and where to rely on human acumen is an important part of deploying AI and HCI systems. The roles and responsibilities of both parties should be aptly defined and coherently distinguished. This research question tries to draw a lucid distinction between these two.
  5. How much, is too much when it comes to the Industry 4.0–HCI confluence?

     Incidentally, as they say, everything is good only in moderation. So, in this section we try to answer an important prerogative on what should be the extent of the Industry 4.0–HCI confluence and what regularization should be enforced at their crossroads, so that the Industry 4.0–HCI ecosystem works in tandem with humans and does not supersede them.

3.4   RESEARCH SOLUTIONS

Complex machinery of the fourth industrial revolution can be quite overwhelming to work with, and with massive chunks of information with respect to each machine, and the interconnection of machines, obtaining the right information at the right time can be quite difficult. Information management and control over such information then become critical to avoid chaos. A carefully designed and well-placed HCI would help to steer clear of such a situation. The meticulous design and information display of an HCI system can be achieved by incorporating design process elements such as qualitative and quantitative analysis, and information architecture. We further stress that this inclusion of AR and VR, would come with a requirement for training to be given to employees working with Industry 4.0 machines, in which case the user interface should be intuitive, and the UX should be a good one. At the same time, employing technologies like AR/VR comes with its very own list of disadvantages. AR/VR headsets are difficult to wear for a long duration of time and would cause an uncomfortable experience. We recommend for implementing the AR/VR correctly into the Industry 4.0 ecosystem, users need to be considered from the start in the design process of these interactions, to understand them and unearth their concerns and pain points through qualitative analysis such as interviews. These concerns should be further validated with a quantitative study such as surveys to realize a hypothesis. This will enable to gauge the value of introducing AR and VR at the initial level, and will also help in making modifications along the way, according to the best-fit scenario with respect to industry personnel.

To answer the aforementioned questions on what recommendations or solutions can be conceived to have a seamless UI–UX component and interaction mechanism and how the manpower–machinery paradigm should work in the HCI, which is specifically developed for and from an Industry 4.0 perspective, we define the following six components of the HCI centered on the Industry 4.0 realm:

As is evident from Figure 3.2, we envision that the UI–UX that is designed keeping industrial recommendations in mind should involve a subset of six core components. AR–VR-based HMDs are the way to go ahead when it comes to industrial supervision roles. It would lessen the need to physically visit the site for inspection and avoid unnecessary hazards. The second component should be the use of Explainable AI models which not only predict and classify but also explain the rationale behind arriving at a specific decision. This is particularly important in the context of Industry 4.0 since there is a significant level of reliance on machines to perform seamlessly day in and out. Thus, knowing why a particular machine took a particular decision at a particular time helps in reasoning and accounts for outlier detections, when the mechanism or machine fails.

Third component of this HCI focuses on the need for a user-feedback network which is pivotal to the smooth functioning of a proper HCI mechanism in

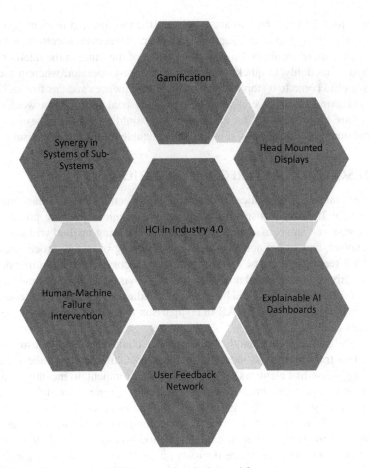

FIGURE 3.2   Components of HCI centered on Industry 4.0.

cyber-physical systems. Users should receive both instantaneous and cumulative feedback for examining fault tolerance and industry-specific bottlenecks in the manufacturing processes, which are often the first deterrents on the assembly line, wherein the entire process comes to a standstill.

Fourth consideration includes the ability to have a combination of HMI when there is a critical failure. Often, there are problems that are only identifiable by human intervention such as system breakdown, actual physical issues with the machinery, leaks, and overflows. During such catastrophic situations, it is important to have a synergy between how much human vs machine intervention is needed for the failure to be resolved in the least possible time, with minimal resources at hand.

The final component of a successful HCI system tailored for the Industry 4.0 framework should include the amalgamation of the system of subsystems which is also known as IoT. In the IoT framework, device, network, and application are the three most important components, which make up the internals in a manufacturing facility. These systems should communicate with each other on a network that is separated from the main hub, to accommodate the passage and relay of commands and messages in a brisk and seamless manner. Whenever a certain subsystem malfunctions, there should be immediate relaying of the same to the interconnected nodes with a modality to quickly intimate the nearest operator, wherein the other systems should come to a stop and avoid dire consequences such as fire and loss of property. Having these six components in an industrial HCI system would enable that there are low downtimes for machinery and would establish a cohesive synergy between humans and machines, through an actionable intuitive mechanism.

3.5   DISCUSSION AND RECOMMENDATIONS

Current modalities and practices of training can become quite mundane and over-whelming for Industry 4.0-specific training. With high-technology procedures to complete tasks on smart machines, the traditional training method will overburden the manpower, and would not be effective in retaining operating procedures. With AR and VR technologies to be used in Industry 4.0 ecosystem, the training provided will be highly engaging and playful, and due to the visual aspects of these technologies, the training imparted would be retained by industry professionals. It could identify the impact of learning with AR in comparison to traditional vocational training approaches.

Similarly, gamification is another modality that could be used in Industry 4.0-specific HCI to train and assist industrial manpower. Gamification is the method of introducing game-like elements into a nongame environment, to incentivize users to do an otherwise mundane task. However, gamification, if not done right, can lead to an even more unpleasant UX than to start with. Therefore, the context of use should be properly defined, and on that front, the concerned manpower should be evaluated in terms of their interest level in game elements. Moreover, stakeholders must be involved in taking an appropriate decision on the type and extent of gamification to be involved in the industry, to benefit the industry as well as personnel mutually (Figure 3.3).

In our humble opinion, an "ideal" Industry 4.0–HCI system should feature a balanced scale of humans and machines working together, with a positive collaboration between them. To achieve such a synergy, humans should not feel excluded from the interconnected smart machines of the industry. Apart from a rational work division between the two entities, for humans to be inclusive in Industry 4.0, high-technology modalities should be used to foster an engaging and exciting experience of co-working between humans and machines. As discussed previously, the intent of using these high-technology modalities, whether they are used to provide information across the industry, or as an assistive technology to complete tasks successfully, needs to be defined clearly and a design structure based on the intent is mandatory to be constructed to gain a first-hand view of the space and function of such technological implementation. At the same time, stakeholders and users must be involved in the design process of Industry 4.0–HCI, to take care of the business requirements and the perspective of users, respectively. For an intelligent Industry 4.0 framework to effectively function with the collaboration of humans and machines, a human-centered HCI design process must be established. This would require a customization

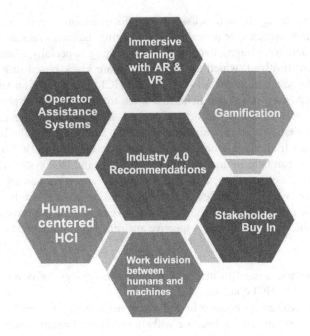

FIGURE 3.3 Industry 4.0–HCI confluence framework for fostering intelligent analytics.

from the traditional HCI design process to fit the characteristics and operational mechanism of the Industry 4.0. Our recommendation is to personalize the process of gathering and understanding requirements, designing solutions based on requirements, testing, and evaluating, in the context of Industry 4.0. The below personalization will cater to the efficient growth of Industry 4.0, all the while maintaining a strong relationship and codependency between humans and machines. To do this we suggest involving domain experts and stakeholders in the user requirement part of the cycle. This will enable the stakeholder groups to minutely specify requirements at the outset, thereby honing the process.

Thereafter, using Explainable Artificial Intelligence (XAI)-based modeling using SHapley Additive exPlanations (SHAP) and Local Interpretable Model-Agnostic Explanations (LIME) model-agnostic frameworks shall help eliminate the black box nature of AI-based solutions and help in bringing fairness and accountability to the process. In the next step, we believe that it is important to not only evaluate the solutions with initial user requirements but also perform usability testing involving stakeholder requirements. This shall help in developing a synergy between users and stakeholders, which shall benefit in the long run by removing the actual constraints faced by users and alleviating them by involving the knowledge of stakeholders. In the last leg of the Industry 4.0–HCI design process cycle, it is important to incorporate user feedback and not only explain them the mechanisms of how the system works but also why it arrived at a certain decision. If there is a large gap between the user-expected output, the stakeholder-predicted output, and the actual output, it is important to reiterate the process by removing critical bottlenecks. This can range from increasing the data set size by conducting more expansive user surveys and user behavior studies, or it can involve changing the train–validate–test dataset sample sizes and percentages, to counter for overfitting, especially when deep learning models are involved, which are inherently more black boxed in nature. Thus, there needs to be a consensus between AI and HCI practitioners on when to draw and derive from their respective fields and when to pause and introspect. The iterative nature of both HCI design processes and AI–ML models make it easier to duly bifurcate and then combine certain tasks with each another. Using the Industry 4.0–HCI confluence framework along with the modifications in the HCI user-centered design process, can help in making the merger of AI and HCI systems transparent, seamless, and cohesive. Thus, after developing the above framework and recommendations for tweaking the user-centered HCI design process, we believe that we can answer the questions posed above. For the sake of reading convenience, we have included the questions before the respective answers using bullet points. They are as follows:

- Why is there a need to involve humans in the process of developing this Industry 4.0–HCI confluence?

  The involvement of users in developing the Industry 4.0–HCI confluence is of utmost importance because without involving domain experts across a wide variety of fields and without involving a randomized, all-encompassing group of users, the solutions would be a black box in nature, like traditional AI models and would severe the purpose of developing human-centered intelligent interactive systems, which depict at least some form and type of intelligence.

- Which stakeholders need to be present in the development of the Industry 4.0–HCI confluence?

  The type and kind of stakeholders which shall be involved in the merger of these two disciplines shall involve AI researchers, factory workers, data scientists, social scientists, philosophers, mathematicians, and other domain experts as per the application area of the respective domain. This involvement is essential to the conducive growth of the field.

- When to trust the AI and when to trust humans, in the decision-making at the Industry 4.0–HCI confluence?

  While it is imperative that users feel empowered when it comes to decision-making, it is equally important for the AI systems to disempower them in a way that they feel less in control of certain processes, which can pose hazards to them. As a very crude and general guideline, if the AI can come to a correct decision repeatedly, with excellent fault tolerance, precision–recall values as well as confusion matrices, the users can learn to trust the AI, especially if Explainable AI toolkits are used to justify and interpret the outputs of the system. All in all, AI systems should be entrusted with tasks and decisions that need huge and fast computation, and for others, humans can be more in control.

- Where can the users work in tandem with the AI/where they cannot, pertaining to the confluence?

Users can work in tandem with AI in the initial design and evaluation phase of the Industry 4.0–HCI confluence and possibly cannot work in certain situations where there is a considerable gap between the respective operational capacities.

- How much, is too much, when it comes to the Industry 4.0–HCI confluence?

AI and HCI systems need to be regularized, made transparent, and trusted with sensitive data that should be kept safe and away from potential attacks. They should also work in tandem with, and for humans, by not overpowering them (Figure 3.4).

In the five-step process represented above, the first and crucial step of designing and setting up Industry 4.0 is to study and analyze the industrial manpower. This is important to keep their experience and concerns in mind while designing intelligent systems and their interfaces around them. The next step involves stating the user requirements analyzed from the qualitative and quantitative study of the concerned manpower and involving every stakeholder to ensure full acceptance of user requirements. Although the above representation only lists stakeholder buy in

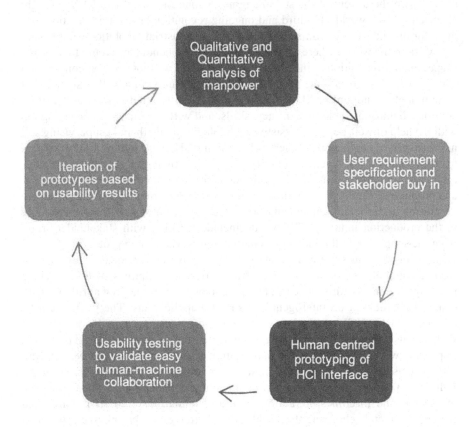

FIGURE 3.4   Intelligent analytical HCI design process for Industry 4.0.

the second stage, it is a practice that should be carried out at each stage. Once the requirements are zeroed upon, prototyping with respect to the complexity of Industry 4.0 machinery and ease of use for manpower should be done. From an Industry 4.0 perspective, it becomes important to make use of a good information architecture and easy-to-use interfaces to operate complex machines. Usability testing on such prototypes reveals inconsistencies and confusion with respect to the usage of inter-faces, on the realization of which, iterations of the prototypes are constructed, and the process is repeated until a good HCI interface is made.

3.6   CONCLUSION AND FUTURE WORK

The development of a HCI system specifically made for the Industry 4.0 environ-ment could be made a reality focusing on the three paradigms discussed in this chapter. It can be a cause of concern for currently working industrial manpower, as it will automate industries extensively. While there are immediate ramifications of this on the numbers of jobs that would be available in the production industry, it does pave the way for developing better industrial practices which are focused on creating a synergy between humans and machines. The first industrial revolution started with the advent of the steam engine, second industrial revolution brought electricity to the world, the third and ongoing revolution brought automation as a boon for the industry and imminently, the fourth industrial revolution would usher a new era in the world, where previously imaginable phenomena would take actual shape. Such a revolution is inevitable, and industries will lose out on competition if they steer clear from Industry 4.0 modifications. But, with a well-designed HCI system in place, not only will the industry workers shift to better roles and respon-sibilities, but also be able to learn new skills, and will experience an engaging job with the help of well-designed assistive technologies to aid them in supervising and maintenance of complex machinery. The aim of an HCI focused on Industry 4.0 is to enable humans and machines to work in tandem with one another and not replace each other, thereby making the Industry 4.0 ecosystem more efficient, transparent, and productive. Future work in this realm would include more case studies that are pertaining to the actual implementation of such cyber-physical systems, especially in the production industry. This would include working with stakeholders right from the shop floor in the industry to the designers who design prototypes for effi-cient cyber-physical system integration with intuitive design ideas. It would also include collaborating between experts from different disciplines such as machine learning, design, cognitive science, and psychology, for developing a truly human-inspired AI focused on intelligent Industry 4.0 applications. The goal would be to synergize AI, HCI, and Industry 4.0 to attain tenable and tangible goals soon, with a focus on improving labor wages, implementing efficient industrial practices, improving work–life balance, and making immediate applications to every sphere in the industry. Thus, a robust framework for implementing HCI systems in the Industry 4.0 framework would open up unending opportunities at the confluence of humans and machines, heralding the era of human-inspired AI research and thereby, ultimately changing the world for the better, one cyber-physical system at a time.

REFERENCES

1. C. Krupitzer et al. A Survey on Human Machine Interaction in Industry 4.0, Vol. 1, No. 1, Article. Association for Computing Machinery (ACM), February 2020. https://doi.org/10.1145/1122445.1122456

2. I. Islam, K. M. Munim, M. N. Islam, and M. M. Karim. "A proposed secure mobile money transfer system for SME in Bangladesh: An industry 4.0 perspective," in *2019 International Conference on Sustainable Technologies for Industry 4.0 (STI)*. IEEE, 2019.

3. J. Lee, B. Bagheri, and H.-A. Kao. "A cyber-physical systems architecture for industry 4.0-based manufacturing systems," *Manufacturing Letters*, vol. 3, pp. 18–23, 2015.

4. H. Gruber. *Innovation, Skills and Investment: A Digital Industrial Policy for Europe, Economia e Politica Industriale*. Springer, 2017, pp. 1–17.

5. C. J. Bartodziej. *The Concept Industry 4+.0: An Empirical Analysis of Technologies and Applications in Production Logistics*. Springer, 2017.

6. J. A. Saucedo-Martínez, et al. "Industry 4.0 framework for management and operations: A review," *Journal of Ambient Intelligence and Humanized Computing*, vol. 6, pp. 1–13, 2017.

7. F. Ansari, S. Erol, and W. Sihn. *Rethinking Human-Machine Learning in Industry 4.0: How Does the Paradigm Shift Treat the Role of Human Learning?* Science Direct, 2018.

8. S. Nahavandi. "Trusted autonomy between humans and robots: Toward human- on-the-loop in robotics and autonomous systems," *IEEE Systems, Man, and Cybernetics Magazine*, vol. 3(1), pp. 10–17, 2017.

9. D. Bauer, S. Schumacher, A. Gust, J. Seidelmann, and T. Bauernhansl. "Characterization of autonomous production by a stage model," *Procedia CIRP*, vol. 81, pp. 192–197, 2019.

10. F. J. Campbell. Human factors: The impact on industry and the environment. In E. R. Rhodes, and H. Naser (Eds.), *Natural Resources Management and Biological Sciences* (pp. 1–14). IntechOpen, 2021.

11. T. Zhang, Q. Li, C. Zhang, H. Liang, P. Li, T. Wang, S. Li, Y. Zhu, and C. Wu. "Current trends in the development of intelligent unmanned autonomous systems," *Frontiers of Information Technology & Electronic Engineering*, vol. 18, no. 1, pp. 68–85, 2017.

12. E. Fosch-Villaronga, C. Lutz, and A. Tamò-Larrieux. "Gathering expert opinions for social robots' ethical, legal, and societal concerns: Findings from four international workshops," *International Journal of Social Robotics*, vol. 12, no. 2, pp. 441–458, 2020.

13. M. Gil, M. Albert, J. Fons, and V. Pelechano. "Designing human-in-the-loop autonomous cyber-physical systems," *International Journal of Human-Computer Studies*, vol. 130, pp. 21–39, 2019.

14. F. Santoni de Sio and J. van den Hoven. "Meaningful human control over autonomous systems: A philosophical account," *Frontiers in Robotics and AI*, vol. 5, pp. 15, 2018.

15. G. Weichhart, A. Ferscha, B. Mutlu, M. Brillinger, K. Diwold, S. Lindstaedt, T. Schreck, and C. Mayr-Dorn. "Human/machine/roboter: Technologies for cognitive processes." *Elektrotechnik Und Informationstechnik*, vol. 136, no. 7, pp. 313–317, 2019.

16. A. M. Isen. "An influence of positive affect on decision making in complex situations: Theoretical issues with practical implications," *Journal of Consumer Psychology*, vol. 11, no. 2, pp. 75–85, 2001.

17. M. N. Islam, S. J. Oishwee, S. Z. Mayem, A. N. Mokarrom, M. A. Razzak, and A. H. Kabir, "Developing a multi-channel military application using interactive dialogue model (IDM)," in *2017 3rd International Conference on Electrical Information and Communication Technology (EICT)*. IEEE, 2017, pp. 1–6.

18. T. Zaki, Z. Sultana, S. A. Rahman, and M. N. Islam, "Exploring and comparing the performance of design methods used for information intensive websites," *MIST International Journal of Science and Technology*, vol. 8, pp. 49–60, 2020.
19. H. Thimbleby and I. H. Witten. "User modeling as machine identification: New design methods for HCI." In H. R. Hartson et al. (Eds.) *Advances in Human-Computer Interaction*. Ablex (vol. 4, pp. 58–86), 1991.
20. A. Beard-Gunter, G. Ellis, C. Dando, and P. Found, Designing industrial user experiences for industry 4.0. 2018.
21. M. A. Razzak and M. N. Islam. "Exploring and evaluating the usability factors for military application: A road map for HCI in military applications," *Human Factors and Mechanical Engineering for Defense and Safety*, vol. 4, no. 1, p. 4, 2020.
22. J. Nielsen. *Usability Engineering*. Elsevier, 1994.
23. S. Deterding, R. Khaled, L. E. Nacke, and D. Dixon. "Gamification: Toward a Definition," in *CHI 2011 Gamification Workshop Proceedings*, vol. 12. Vancouver, BC, Canada, 2011.
24. B. Shneiderman, C. Plaisant, M.S. Cohen, S. Jacobs, N. Elmqvist, and N. Diakopoulos. *Designing the User Interface: Strategies for Effective Human-Computer Interaction*. Pearson, 2016.
25. S. Uzor, J.T. Jacques, J.J. Dudley, and P.O. Kristensson. "Investigating the accessibility of crowdwork tasks on mechanical turk," in *Proceedings of the 2021 CHI Conference on Human Factors in Computing 2021*, https://doi.org/10.1145/3411764.3445291
26. Sebastian Büttner, Henrik Mucha, Markus Funk, Thomas Kosch, Mario Aehnelt, Sebastian Robert, and Carsten Röcker. "The design space of augmented and virtual reality applications for assistive environments in manufacturing: A visual approach," *ResearchGate Conference Paper*, June 2017.
27. Markus Funk, Albrecht Schmidt, Tilman Dingler, and Jennifer Cooper. "Stop Helping Me - I'm Bored! Why Assembly Assistance needs to be Adaptive," *ResearchGate*, September 2015.
28. D. Gorecky, M. Schmitt, M. Loskyll, and D. Zuhlke, "Human-Machine- ˝ Interaction in the Industry 4.0 era," in *2014 12th IEEE International Conference on Industrial Informatics (INDIN)*. 2014, pp. 289–294.
29. E. Aranburu, G. S. Mondragon, G. Lasa, J. K. Gerrikagoitia, I. S. Coop, and G. S. Elgoibar. "Evaluating the human machine interface experience in industrial workplaces," in *Proceedings of the 32nd International BCS Human Computer Interaction Conference*. BCS Learning & Development Ltd., 2018, p. 93.
30. Peter Papcun, Erik Kajáti, and Jií Koziorek. *Human Machine Interface in Concept of Industry 4.0*. IEEE, October 2018.
31. R. Tulloch. Reconceptualising gamification: Play and pedagogy. Online.Available: http://www.digitalcultureandeducation.com/cms/wpcontent/uploads/2014/12/tulloch.pdf.
32. T. Schmidt, I. Schmidt, and P. R. Schmidt. "Digitales Spielen und Lernen – A Perfect Match?: Pädagogische Betrachtungen vom kindlichen Spiel zum digitalen Lernspiel," In K. Dadaczynski, S. Schiemann, and P. Paulus, (Eds.) *Gesundheit spielend fördern: Potenziale und Herausforderungen von digitalen Spieleanwendungen für die Gesundheitsförderung und Prävention*, 2016, pp. 18–49.
33. Jacqueline Schuldt, and Susanne Friedemann. "The challenges of gamification in the age of Industry 4.0." in *IEEE Global Engineering Education Conference*, April 2017.
34. T. Schmidt. MALL meets Gamification: Möglichkeiten und Grenzen neuer (digitaler) Zugänge zum Fremdsprachenlernen. http://www.uni- potsdam.de/fileadmin01/projects/tefl/
35. Sebastian Büttner, Henrik Mucha, Markus Funk, Thomas Kosch, Mario Aehnelt, Sebastian Robert, and Carsten Röcker. "The design space of augmented and virtual reality applications for assistive environments in manufacturing: A visual approach," in

Proceedings of the 10th International Conference on Pervasive Technologies Related to Assistive Environments - PETRA '17. ACM Press, Island of Rhodes, Greece, 2017, pp. 433–440. https://doi.org/10.1145/3056540.3076193

36. G. Dini and M. Dalle Mura. "Application of augmented reality techniques in through-life engineering services," *Procedia CIRP*, vol. 38, pp. 14–23, 2015. https://doi.org/10.1016/J.PROCIR.2015.07.044

37. C. D. Fehling, A. Mueller, and M. Aehnelt. "Enhancing vocational training with augmented reality." In: S. Lindstaedt, T. Ley, and H. Sack (Eds.) *Proceedings of the 16th International Conference on Knowledge Technologies and Data-Driven Business.* ACM Press, 2016.

38. S. Deterding, D. Dixon, R. Khaled, and L. E. Nacke, *Gamification: Toward a Definition.* CHI, 2011.

39. SangSu Choi, Kiwook Jung, and Sang Do Noh. "Virtual reality applications in manufacturing industries: Past research, present findings, and future directions," *Concurrent Engineering: Research and Applications*, vol. 23, no. 1, pp. 40–63, 2015. https://doi.org/10.1177/1063293X14568814

40. R. Palmarini, J. Ahmet Erkoyuncu, and R. Roy. An innovative process to select augmented reality (AR) technology for maintenance. *Procedia CIRP*, vol. 59, pp. 23–28, 2017. https://doi.org/10.1016/J.PROCIR.2016

41. M. Hermann, T. Pentek, and B. Otto. "Design principles for Industrie 4.0 scenarios, 2016," in *49th Hawaii International Conference on System Sciences (HICSS).* IEEE, 2016, pp. 3928–3937. https://doi.org/10.1109/HICSS.2016.488

42. F. Hecklau, M. Galeitzke, S. Flachs, and H. Kohl. Holistic approach for human resource management in Industry 4.0. *Procedia CIRP*, vol. 54, pp. 1–6, 2016. https://doi.org/10.1016/J.PROCIR.2016.05.102

43. S. Zahoor, M. Bedekar, and V. Vishwarupe. A framework to infer webpage relevancy for a user. In S. Satapathy and S. Das (eds) *Proceedings of First International Conference on Information and Communication Technology for Intelligent Systems: Smart Innovation, Systems and Technologies*, vol 50. Springer, Cham, 2016. https://doi.org/10.1007/978-3-319-30933-0_16

44. S. Zahoor, M. Bedekar, V. Mane, and V. Vishwarupe. Uniqueness in User Behavior While Using the Web. In: S. Satapathy, Y. Bhatt, A. Joshi, and D. Mishra (Eds.) *Proceedings of the International Congress on Information and Communication Technology. Advances in Intelligent Systems and Computing*, vol. 438. Springer, Singapore, 2016. https://doi.org/10.1007/978-981-10-0767-5_24

45. V. Vishwarupe, M. Bedekar, and S. Zahoor. "Zone specific weather monitoring system using crowdsourcing and telecom infrastructure," in *2015 International Conference on Information Processing (ICIP)*, 2015, pp. 823–827, doi: 10.1109/INFOP.2015.7489495

46. V. Vishwarupe, M. Bedekar, M. Pande, and A. Hiwale. Intelligent Twitter spam detection: A hybrid approach. In: X. S. Yang, A. Nagar, and A. Joshi (Eds.) *Smart Trends in Systems, Security and Sustainability. Lecture Notes in Networks and Systems*, vol. 18. Springer, Singapore, 2018. https://doi.org/10.1007/978-981-10-6916-1_17

4 Advance in Robotics Industry 4.0

Muchharla Suresh, Rajesh Kumar Dash,
Deepika Padhy, Swastid Dash,
Subhransu Kumar Das, and
K. Nageswara Rao Achary
NIST

CONTENTS

4.1 INTRODUCTION

Due to the advent and prominence of Industry 4.0, robotic technology, which provided an essential contribution to modern industry, has undergone significant development in recent years. In the era of Industry 4.0 and the Industrial Internet of Things (IIoT), it is anticipated that the next age of robotics and its related technologies will play a major role in meeting the flexible needs of cooperative and smarter production [1]. Industrial robots are embedded with artificial intelligence (AI), which are capable of replacing the human workers in the manufacturing industries. Human labor is ineffective in factories for a variety of reasons, including physical limitations that affect the production performance, manufacturing costs, etc. In the current competitive business, firms have turned away from using human labor and toward industrial robots in order to accomplish more accurate production in shorter amounts of time without even any risks [2].

However, for effective and reliable manufacturing, human and robot collaboration is necessary. Industrial robots are necessary for companies in the modern, tough marketplace not just for safety reasons to prevent worker injuries while production as well as for the most need faster, more precise manufacturing when taking financial prosperity into account. Real-time result and effectiveness predictability in manufacturing do not aid in the unsupervised management and optimization of the subsystem cost and duration. The phrase "Industry 4.0" refers to a new industrialization that is

DOI: 10.1201/9781003321149-4

built on smart technology and intends to integrate data science into the manufacturing and production processes to create smart factories. The robots can also independently spot a decline in product performance and apply optimization to fix it [3].

Industrial robots are created extremely effectively and cooperatively with people and other robots through networking in the Industrial revolution 4.0, enabling them to be self-adaptable and self-aware on new goods and production techniques. As a result, smart factories with many of these robots will be provided in the future of industry thanks to recent technologies utilizing IoT, such as for controlling and wirelessly monitoring on robots, cloud platforms analyzing big data has enhanced information insights. The robots may also detect product performance degradation on their own and apply modifications to fix it [1].

About 373,000 industrial robots had been based on these factors in 2019, according to the International Federation of Robotics (IRF). Globally, there were 2.7 million industrial robots in use in factories in 2020. The effective use of industrial robots, their dependability and affordability, and the effective adoption of the Industry 4.0 concept have sparked an increase in interest in robot enhancement and the investigation of latest advancements in a variety of fields, especially in typical and nontypical application domains. Using the term "Industrial robot" as a keyword, more than 4,500 scientific papers had been published in 2019, and 5,300 papers with a similar interest and study focus had been published in 2020, according to Science Direct, one of the largest research journals. Regardless of the legislative, socioeconomic, and sociological forces influencing the market for new robots, scientific interest in this topic is based on a continual increase in publications [4].

This scholarly research is primarily concerned with the industrial robot applications in 2018–2021 in industries, where there has historically been little support

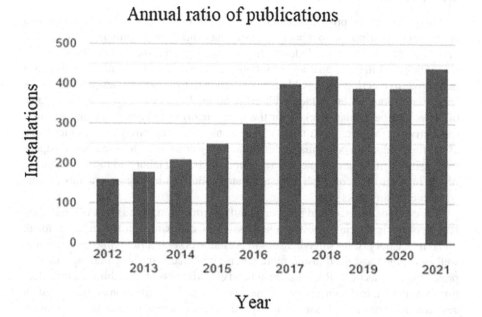

FIGURE 4.1 Annual ratio of publications.

Top 10 markets of industrial robotics

Rank	Country	Units Sold	%Change
1	China	87,000 Units	58%
2	Korea	41,400 Units	4%
3	Japan	38,600 Units	18%
4	United States	31,400 Units	6%
5	Germany	20,000 Units	8%
6	Taiwan	7,600 Units	44%
7	Italy	6,500 Units	19%
8	Mexico	5,900 Units	7%
9	France	4,200 Units	16%
10	Spain	3,900 Units	No data

* Comment

6/27/22, 10:00 PM   Table Chart Maker Online

Project Market Size Of Industrial Robotics

Rank	Country	Units Sold
1	China	170,000 Units
2	Japan	45,000 Units
3	United States	45,000 Units
4	Korea	44,000 Units
5	Germany	23,500 Units
6	Taiwan	12,000 Units
7	Italy	7,500 Units
8	Mexico	7,000 Units
9	Spain	5,000 Units
10	France	5,100 Units

* Comment

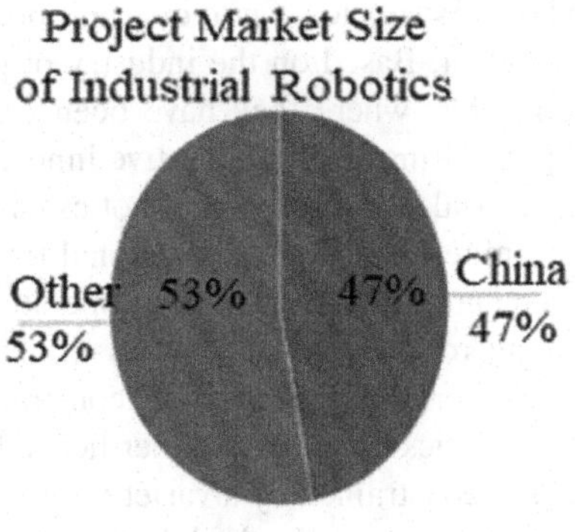

Project Market Size
of Industrial Robotics

Other 53%   47% China
53%   47%

FIGURE 4.2 Markets of industrial robots.

for robotization. It also covers core topics including feature extraction, route planning, and optimizations, as well as human contact. Although the field of regular automation and general societal widespread usage of smart gadgets, semi uses of robotics are still frequently viewed with significant suspicion. The most pervasive misconception regarding robots is that they will take over human jobs, depriving human employees without the need for a means of support [5].

Figure 4.1 indicates the yearly ratio of implemented robot publications in industries. Here, the number of installations of industrial robots is increasing year by year. The idea was first introduced in 2012, and still, this process is going on with the addition of new features and with new technologies.

However, the research presented in, which evaluated the huge backlash about robots replacing humans in Japan's fabric and electronics industries, disproved this notion. Based on an analysis of the number of robots used and the actual cost of implementation, it was found that the employment of robots increases productivity, which benefits the much more vulnerable individuals in society, including women, part-time or full-time workers, college grads, and elderly people. Every robotization task is distinctive in some manner. In order to increase the functionality or enhance the qualities of standard automatons, these tasks frequently call for the use of specific tools, the development of an appropriate working environment, the addition of new sensors or measurement systems, and the application of sophisticated control algorithms. Industrial robots typically work in larger groups called robotic units or independent production lines. As a consequence, especially seemingly simple tasks that are robotized require a complicated response and a structured method. Figure 4.2 depicts the market of industrial robots.

4.2   OVERVIEW OF INDUSTRY 4.0

Due to the rise of Industry 4.0 and its widespread use, robotic innovations provided an essential contribution to contemporary manufacturing and have undergone significant growth in recent times. According to predictions, the next wave industrial robots and the innovations that go along with it will play a bigger part in addressing the changing demands of collaborated and interactive production within the framework of Industry 4.0 [6]. The actual use of a robotic system is indeed a challenging task that demands providing answers to several queries upon the feasibility of doing so and the procedure altogether. Based on the industry of production, the situation differs a little. In certain fields, where they have been used for at least 30 years, robot implementation is just getting started. Creative innovations are comparatively easier to implement in industrial settings where robotics has a lengthy history. Such approaches are generally limited about using additional tools, computational methods, and process improvement systems for robotic operation. Although their wide range of applications, many robotics advancements might effectively move from one sector to the other. Furthermore, overcoming constraints in one area secures improvements throughout robotics in plenty of other fields. Industrial robots' ability to complete common tasks is constrained by a variety of factors. Great identification efficiency demands, absence of adequate methodologies, and functional requirements are the key constraints. Except for the tools, these limitations have nothing to do with the mechanical systems of the robot. Thus, the development of new machinery or optimization techniques stimulates the majority of mainstream robotic applications. Machine learning (ML), sensor technology, and data analysis are arguably the main forces behind the expanded use of industrial robots. Despite the fact that robots primarily intended to do repetitive tasks, technologies create robots extremely adaptable as well as enable them to come up with clever responsive alternatives. Internal structures of robots have improved as a result of the development of robot control systems. These improvements frequently involve the use of new computational models enabling the control of robots [7].

4.3 OPPORTUNITIES OF INDUSTRY 4.0

Zero downtime and maximum efficiency are the goals of Industry 4.0-enabled robotics [8]. Robotics is an essential Industry 4.0 technology that offers a wide range of capabilities in the industrial area. Automation systems have been improved because of this technology. Industry 4.0, often known as the factory of the future, employs sophisticated robots to boost efficiency by taking over manual jobs and completing them quicker. This technology performs repeated tasks with greater accuracy and at a reduced cost. One example is the assembling of flexible pieces. Because robots can operate in hazardous areas, they eliminate risky employment for people. They are capable of carrying huge items, hazardous material, and doing repeated jobs. This has aided businesses in preventing several mishaps while also saving time and money. Recent advancements in industry 4.0, the Internet of Things, and AI are expected to have a significant influence on robotics. New maintenance tactics, more autonomous robots, and new collaborative robotics technologies are made possible by Industry 4.0 technology.

Industrial robotics may execute many functions at the same time. By replacing a massive multicore processing capability with an existing programmable logic controller, facility administrators might efficiently leverage industrial floor space while shrinking the hardware footprint (Programmable Logic Controller, PLC). Major food firms have integrated robots into their existing food production operations. With vision technologies, cameras, and AI, the robots perform a variety of tasks in production. Everything is being cut, measured, packed, and palletized by them. Industries with an abundance of sensors will track machinery and manufacturing processes in real time to forestall anomalous output and services. Machine vision robots are able to execute sophisticated optical procedures with pinpoint accuracy. Microscopic flaws or slight color variations can be recognized and repaired instantly to maintain performance quality [4].

Automation is becoming far more trustworthy than human labor. For decades, the business has used robots on floors. The robots are equipped with lasers and cameras, allowing for high-precision welding. Because of the effectiveness of automation, manufacturers may eliminate overall waste in the manufacturing chain [5]. When robotic systems may be improved or reallocated as the market model evolves to fulfill new operations, replacement expenses can be reduced. Although they may appear to be a hefty initial investment, industrial robots rapidly pay for themselves through decreased labor costs and quicker production cycles. Long-term operating and service expenses would be cheaper than if the same functions were performed by an employee [9–11].

The installation of a smart camera at the end of a robotic arm opens up a wide range of applications since the arm will traverse over the investigated component to monitor numerous characteristics. Computer vision uses a variety of innovations to deliver realistic outcomes from picture gathering and interpretation for robot-based surveillance and assistance. Vision system inspection, which is rapidly expanding, is used to identify surface defects, color, and existence. The vehicle sector has been a significant driver of industrial robotics, using the majority of the robots now are in use. Body welding has been the most prominent robotic activity in vehicle manufacture. At one point during this process, two metal pieces are fused together to generate a combined fusion of both materials.

Robots that are programmed to perform continuously throughout the day. In practically any condition to assist industrial organizations in increasing production, throughput, and profitability. Despite the fact that automation disrupts the core of employment, it has offered significant profit prospects in the industrial business. One of the most frequent types of automation robots is the collaborative robot. Collaborative robotics will become more versatile in manufacturing and perform creative activities as technology evolves. Alternatively, recent sensor designs and algorithms have evolved to guarantee that early robots are more suited for big quantities, minimum variations, and novel purposes. Robots are so rapidly being employed in a wide range of industries, including manufacturing [6,7].

Robots are widely employed to apply sealing material cables or adhesives in a variety of businesses. The material to be applied is liquid or paste in consistency. In addition, robots are employed in the automobile sector to drill, install, identify, modify, and process improvements. By automated production, industrial robots assist many manufacturing industries in increasing efficiency and improving performance. Sorting, wrapping, and packaging are all significant jobs. Work that is labor intensive and time-consuming may be done concurrently. It is unrealistic to expect mankind to have unlimited energy reserves. Furthermore, people will make blunders while completing projects [12,13].

4.4 CHALLENGES IN INDUSTRIAL ROBOTICS 4.0

Industrial robots are increasingly being connected to the cloud. This means they can gather data about themselves to enhance productivity, service standards, and security, all while lowering the overall cost. In the coming decades, industrial robot developers will encounter a slew of other new challenges. As these systems get more complicated, the ability to guarantee reliability becomes extremely critical. Its purpose is to transform existing industrial production plants into smart factories of the future by incorporating various technologies such as the IoT, cloud computing, and AI.

Industrial robots use a variety of memory methods, including storing the boot code as expansion memory, associated with data memories. With the introduction of Industry 4.0, there is a strong need to secure systems from cyber attacks.

Because of the possibility of injury, human–robot contact is now challenging. Manufacturers prior to Industry 4.0 dealt with the problem by not allowing humans and robots to share workstations. This will change when superior AI develops, allowing these people and a new generation of cooperative robots to work together. Collaboration entails people and robots working closely together while maintaining flexibility and productivity. People and robots will be able to work more closely together while maintaining flexibility and productivity, says Microsoft's artificial intelligence boss, Satya Nadella in an interview with BBC World News.

A new type of cobot is also forming. Chatbots are computer programs that are designed to assist knowledge workers by gathering data from backend systems, such as current inventory levels or shipment arrival times. They will work with a range of interfaces, including the web, smartphone apps, and augmented reality glasses.

4.5 PROPERTIES OF ROBOTICS INDUSTRY 4.0

The industrial manufacturing developed speedily in the last few years. Ten years ago, the German government began encouraging manufacturers to exploit new technologies and embrace digitalization. The Industry 4.0 initiative, as it becomes known, appeared as the early concepts of the IoT started to gather momentum. Industrial potential for IoT was timely, and the IIoT formed a symbiotic and synonymous relationship with Industry 4.0. The widespread use of industrial robots started decades before. However, together the IIoT and Industry 4.0 enabled factory-wide machinery connectivity, performance monitoring, and predictive asset maintenance on a scale not previously possible. The abilities of industrial robots have advanced considerably, from single-axis, single-function units to more complex multiaxis robots that have interchangeable grippers, manipulators, and effectors. Today's industrial robots are able to withstand the harsh and physically demanding environments that are experienced across the many different manufacturing processes. Their ability to continually perform repetitive and complex tasks precisely improves the overall operating effectiveness of any manufacturing function significantly. As robotic systems and electronics technology evolved, researchers started investigating how programmers could train robots to work collaboratively alongside skilled human workers. The concept of a collaborative robot, or "cobot", was born and is gathering momentum [4] with its applications as follows:

  i. **Technologies for Better Vision, Localization, and Supervision**
     Instead of operating in solitary surroundings, robotic systems may operate independently as well as effectively among people in regard to powerful output detection as well as visual technologies.
 ii. **Making Use Out of Computer Vision through Robot Control Systems**
     The ability of the robotic controllers to be connected to a virtual reality tool allows the operator to see the motives, locations, territories target area, and whether as has been demonstrated throughout publications also in a number among the most recent years of advertisement robots. According to the above, these can be employed when channels of communication are restricted, such as when wearing rubber gloves prevents the use of a coaching necklace, as well as when boundaries are present. As a result, fully functioning human sustainability as well as operability may be implemented at the community level without needing to work in the automation field [12,13]. In addition to enabling collaborative information exchange

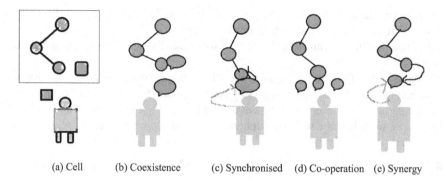

(a) Cell   (b) Coexistence   (c) Synchronised   (d) Co-operation   (e) Synergy

FIGURE 4.3   Basic level of collaboration.

between humans and robots, this capability also prevents robots from doing their duties autonomously.

iii. Technologies for Such IIoT and Wireless Transmission Potential

The most notable technology in today's industrial generation is arguably the IoT. IoT investment is anticipated to reach its peak in 2023 at $450 billion [14]. In 2017, ABB Robotics debuted its own "ability" technology. Such technologies were created to achieve huge manufacturing with something like a high level of real-time customization adaptability [15].

Further researches revealed how cloud computing and robots may work together for commercial applications and how robots improved performance that encourages the implementation of industry 4.0.

iv. Fundamental Robotization Techniques

In industrial robotics, there are five standard degrees of human–robot interaction, including (i) no collaboration, (ii) coexistence, (iii) synchronization, (iv) cooperation, and (v) collaboration.

The basic level of collaboration is covered by the classical strategy as shown in Figure 4.3. It is based on the strategies that robots use to keep people out of their workspace, such as creating sealed robotic compartments where human activity is forbidden; if a person was to enter the robotic work area, the robot would need to be halted. The above method denotes the employment of multiple security measures to identify and prohibit human access to the robotic work area. The final four collaboration degrees make up the modern strategy. This is mostly based on an alternative viewpoint that contends that humans and robots may coexist and operate together in the same workplace. Such strategies generate new requirements for the architecture, management, and detection systems of robots. Machines and robots are often known as cobots, which are designed to operate alongside the human workforce [16].

4.6   CONCLUSION

To help companies fulfill expanding customer demand and maintain their competitiveness in the global market, industrial manufacturing is continually changing.

Currently, robotic devices are penetrating a number of industrial markets. Robotic gadgets will soon be accessible to customers in a variety of formats as they become more reasonably priced, now with the potential to have a variety of effects on our lives. Applications of robotics in the industrial industry have increased protection, quality, and sustainability for businesses. Robotics integration is advantageous to the industrial sector with the arrival of Industry 4.0 for a number of reasons, including dependability, accuracy, efficiency, and tolerance to harsh situations. It could be applied to Industry 4.0 to enable more logical decision-making. Additionally, it can be used in conjunction with business procedures to improve collaboration among different data systems. It offers a production method that is more dependable and efficient. Intelligent robots that work quickly and precisely have improved several production processes in the sector. Due to the current demand, solutions that enable frequent product modifications at a cheap cost are required. As a consequence, factory robots have become the best choice for automated assembly. Automation will therefore provide the manufacturing sector with significant profit prospects in the future years [17].

REFERENCES

1. L. S. Dalenogare, G. B. Benitez, N. F. Ayala, and A. G. Frank, "The expected contribution of Industry 4.0 technologies for industrial performance," *Int. J. Prod. Econ.*, vol. 204, pp. 383–394, 2018.
2. G. Adamson, L. Wang, and P. Moore, "Feature-based control and information framework for adaptive and distributed manufacturing in cyber physical systems," *J. Manuf. Syst.*, vol. 43, pp. 305–315, 2017.
3. R. Dekle, "Robots and industrial labor: Evidence from Japan," *J. Jpn. Int. Econ.*, vol. 58, p. 101108, 2020.
4. K. Aggarwal, S. K. Singh, M. Chopra, S. Kumar, and F. Colace, "Deep learning in robotics for strengthening Industry 4.0.: Opportunities, challenges and future directions," in N. Nedjah, A. A. Abd El-Latif, B. B. Gupta, and L. M. Mourelle (eds) *Robotics and AI for Cybersecurity and Critical Infrastructure in Smart Cities*, Springer Cham, 2022, pp. 1–19.
5. E. Erős, M. Dahl, A. Hanna, P.-L. Götvall, P. Falkman, and K. Bengtsson, "Development of an industry 4.0 demonstrator using sequence planner and ROS2," in Anis Koubaa (ed) *Robot Operating System (ROS)*, Springer, 2021, pp. 3–29.
6. B. Kehoe, S. Patil, P. Abbeel, and K. Goldberg, "A survey of research on cloud robotics and automation," *IEEE Trans. Autom. Sci. Eng.*, vol. 12, no. 2, pp. 398–409, 2015.
7. I. Zambon, M. Cecchini, G. Egidi, M. G. Saporito, and A. Colantoni, "Revolution 4.0: Industry vs. agriculture in a future development for SMEs," *Processes*, vol. 7, no. 1, p. 36, 2019.
8. A. Olivares-Alarcos, S. Foix, and G. Alenya, "On inferring intentions in shared tasks for industrial collaborative robots," *Electronics*, vol. 8, no. 11, p. 1306, 2019.
9. A. M. Ghouri, V. Mani, Z. Jiao, V. G. Venkatesh, Y. Shi, and S. S. Kamble, "An empirical study of real-time information-receiving using industry 4.0 technologies in downstream operations," *Technol. Forecast. Soc. Change*, vol. 165, p. 120551, 2021.
10. T. Brogårdh, "Present and future robot control development—An industrial perspective," *Annu. Rev. Control*, vol. 31, no. 1, pp. 69–79, 2007.
11. V. Villani, F. Pini, F. Leali, and C. Secchi, "Survey on human–robot collaboration in industrial settings: Safety, intuitive interfaces and applications," *Mechatronics*, vol. 55, pp. 248–266, 2018.

12. F. D. Ferreira, J. Faria, A. Azevedo, and A. L. Marques, "Product lifecycle management enabled by industry 4.0 technology," *Transdiscipl. Eng.*, vol. 6, no. 3, pp. 349–354, 2016.
13. A. T. Rizvi, A. Haleem, S. Bahl, and M. Javaid, "Artificial intelligence (AI) and its applications in Indian manufacturing: A review," in S. K. Acharya, D. P. Mishra (eds) *Current Advances in Mechanical Engineering. Lecture Notes in Mechanical Engineering.* Springer, Singapore, 2021. https://doi.org/10.1007/978-981-33-4795-3_76.
14. B. Khoshnevis, "Automated construction by contour crafting—Related robotics and information technologies," *Autom. Constr.*, vol. 13, no. 1, pp. 5–19, 2004.
15. I. Malý, D. Sedláček, and P. Leitao, "Augmented reality experiments with industrial robot in industry 4.0 environment," in *2016 IEEE 14th International Conference on Industrial Informatics (INDIN)*, 2016, pp. 176–181.
16. E. Ruffaldi, F. Brizzi, F. Tecchia, and S. Bacinelli, "Third point of view augmented reality for robot intentions visualization," in *International Conference on Augmented Reality, Virtual Reality and Computer Graphics*, 2016, pp. 471–478.
17. B. Insider, "Business insider intelligence report." 2018.

5 A Cloud-Based Real-Time Healthcare Monitoring System for CVD Patients

Anjali Kumari, Animesh Kumar,
Kanishka Vijayvargiya, Piyush Goyal,
Ujjwal Tripathi, and Avinash Chandra Pandey
PDPM IIITDM

CONTENTS

5.1 INTRODUCTION

Quality healthcare is a need of this century. Chronic diseases such as heart disease, cancer, and diabetes have risen dramatically due to economic, environmental, and social development and lifestyle changes. Quality in health care is essential and refers to ensuring that the patient receives the proper care at the right time. Delays in monitoring a patient might lead to adverse effects and can even be fatal sometimes. An inefficient healthcare system adds up to poor safety, poor care coordination, and inefficiencies, costing millions of lives. The healthcare system in India is not well regulated. Rural areas suffer the most from unattended health checkups. According to a report published by Novartis, Arogya Parivar: Health for the Poor, about 72% of the elderly population live in villages [1]. This is often an alarming number considering that the population of rural areas does not seem to be attended to too well by health officials. Therefore, it is important to encourage reforms that result in health systems that are more resilient, better centered on the needs of people, and sustainable over time. Thus, the idea is to implement systems that make it easier for doctors to understand the areas of abnormality and attend to the sick in their homes. If the health condition passes a specific threshold, and with the assistance of machine learning (ML) analysis, it becomes easier to investigate the world as a whole and deploy any resources if necessary. We use artificial intelligence (AI) due to its significant impact on recognizing and predicting patient health status based on physiological indications. These readings can be acquired using a variety of biological signal-measuring instruments. We introduced a brand-new cloud-based, real-time health monitoring system in this chapter that makes use of Internet of Things (IoT) and AI. The proposed system can analyze real-time biological data and predict the health condition of the user. The proposed system also recommends possible treatments for the users.

The rest of the chapter is arranged as follows: Section 5.2 discusses the state-of-the-art approaches for real-time health monitoring, while in Section 5.3, machine learning classifiers have been discussed. The proposed model has been discussed in Section 5.4, and Section 5.5 assesses the efficiency of the proposed method. In Section 5.5, the application areas of the proposed model have been discussed, and Section 5.7 concludes the chapter.

5.2   RELATED WORKS

Many researchers have worked in a wide range of medical sectors in order to develop a more effective health monitoring system that would help people in maintaining their health. Almost all other countries have been dealing with this public health issue for the past 10–15 years. There have been numerous attempts by various research groups to develop effective regression and classification models using machine learning and deep learning methodologies to predict health-related risks. The well-known classifiers logistic regression, random forest, multiplayer perception, and Gaussian Naive Bayes were used to create high voltage classification using electroencephalogram (EEG) signals. Several methods have been presented using AI, IoT, big data, virtual reality, 3D scanning, 3D printing, and cloud computing [2]. Bhanuteja et al. [3] discussed the various applications of Industry 4.0 technologies and also identified the requirements of customized medical gear. Besides, they collected information for healthcare systems to treat COVID-19 patients properly. Ahmad et al. [4] presented CAMISA, an AI-powered system for remote monitoring of COVID-19 patients. The proposed system creates sensor device in the form of a smart shirt and a nebulizer equipped with several sensors for obtaining a patient's physiological information. The system measures a patient's pulse, SpO2 level, temperature, and breathing rate. If these metrics surpass their threshold values, notifications are sent to the hospital, including the patient's current location. In addition, the authors have used a neural network model to estimate the likelihood of a user contracting a coronavirus. Palwe et al. [5] proposed an enhanced CNN model to identify the suspicious person suffering from COVID-19. Habib et al. [6] discussed many IoT-based approaches presented to handle various health monitoring system. Further, they also discussed the usage of IoT method to detect COVID-19. Chern et al. [7] presented an ensemble hard voting model for cardiovascular disease prediction. They also observed that the advances in machine learning technologies have enabled early disease identification and diagnosis

5.3 PRELIMINARY WORK

A supervised machine learning algorithm understands the function that transforms an input into an output. It learns this function from analyzing the training data and can then produce outputs for unknown inputs, which is explained as follows:

i. Decision Tree Classifier (DTC): DT is a paradigm for supervised machine learning [7]. In the sense that each node (representing one property) links into two or more subnodes, it can be compared to or thought of as a flow-chart. A DTC predicts the value of a variable using the decision rules it has learned or inferred from the training data, like how flowcharts support decision-making processes by adhering to basic assertions.

ii. Random Forest Classifier (RFC): RFC is a machine learning model that is used for both classification and regression and is an ensemble model [8]. To obtain a greater degree of accuracy in prediction, RFCs merge numerous decision trees.

iii. Logistic Regression (LR): a discrete output's likelihood can be modeled using the LR method given an input variable [9]. The most common logistic regression models produce binary results, which might be true or false, yes or no, and so on.

iv. Gaussian Naive Bayes (GNB) Classifier: GNB is a classification method based on Bayes' theorem and the assumption of predictor independence [10]. A Naive Bayes classifier believes that the presence of one feature in a class has no influence on the existence of subsequent features.

v. Ensemble Learning: a broader meta-approach to machine learning called "ensemble learning" aims to improve predictive performance by aggregating predictions from many models [11]. Ensemble learning is primarily used to enhance a model's performance (classification, prediction, function approximation, etc.) or lessen the possibility of making a mistaken choice of a subpar model.

vi. Bagging: bagging, also known as bootstrap aggregation, is an ensemble learning technique that is widely used to reduce variance in noisy datasets. A single data point may be chosen more than once during bagging, which chooses a random sample of data from a training set with replacement [12]. After the creation of numerous data samples, these weak models are individually trained, and depending on the task—for example, classification or regression—the average or majority of those predictions leads to a more accurate estimate.

5.4 METHODOLOGY

This chapter introduces a novel real-time health monitoring system using IoT and various ML approaches. The flow of the proposed approach has been presented in Figure 5.1. The proposed IoT-based remote monitoring combines the capabilities of IoT and ML to predict areas that are more susceptible to any rise in diseases. The proposed ML model is first trained with ground truth and then it can be used for analyzing the real-time data. The real-time data are collected by biological sensors and

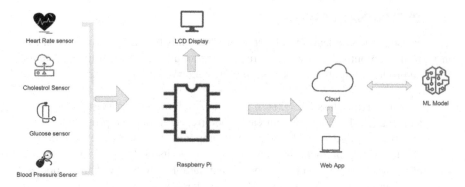

FIGURE 5.1   Proposed methodology.

hosted on the cloud to keep track of the records of patients, which in turn improves the model. Moreover, a notification system has also been implemented, which will trigger an alert in critical conditions to medical professionals and caretakers of the patient.

Further, the overall architecture that shows the tools used for developing the complete arrangement is depicted in Figure 5.2. The overall architecture is based on healthcare IoT, which administers the physiological factors via the vital signs affecting a patient's health. All the sensors are interfaced with the Raspberry Pi, which is a microprocessor. The developed hardware enables the system to collect and transfer vital sign data to cloud storage. The vital signs to be measured with this design are heart rate, blood pressure, oxygen saturation (SpO2), and body temperature. These sensed parameters are sent to the edge computing device (Raspberry Pi). Besides, the detailed description of each phase to monitor the health condition of elderly people has been discussed in the following section.

First, the data acquisition phase is used to collect real-time data of the target user using a biological sensor, and then collected data are preprocessed to make it ready for the analysis. After that, ML models are used to detect any abnormality in real time and possible treatments are also recommended.

5.4.1   DATA ACQUISITION

The first step is to create a dataset that will hold the training data for the model. Here, we have used a cardiovascular disease dataset from Kaggle, which includes parameters such as age, height, weight, gender, systolic blood pressure, diastolic blood pressure, cholesterol, glucose, smoking, alcohol intake, and physical activity to predict the presence or absence of cardiovascular disease.

5.4.2   DATA PREPROCESSING

The collected data are cleaned and prepared for model training. If any parameters contain null values or outliers, those values are dropped from the dataset. Highly

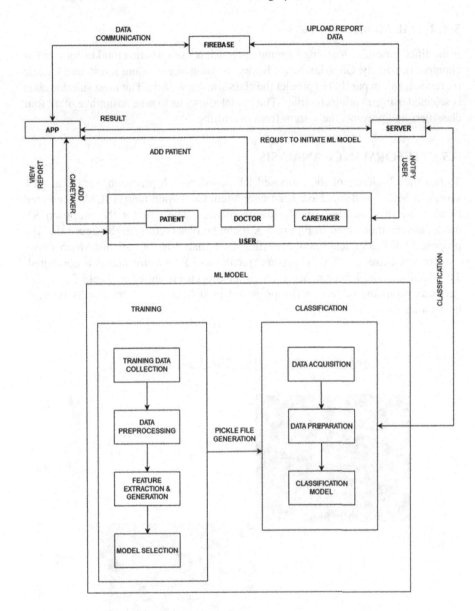

FIGURE 5.2 Overall architecture.

correlated features are used to create new features, like weight and height are used to create BMI. All features are quantized into bins and further encoded using label encoding techniques. As gender has an enormous impact on the presence of cardiovascular disease, the dataset is divided into two, and then clustering analysis (KModes clustering) is performed. The predicted clusters are added as features to the new combined dataset on which feature selection is performed. The prepared dataset is divided into three parts for training, validating, and testing.

5.4.3 ML MODELS

A modified ensemble learning bagging approach is used where a total of four different models, namely, Gaussian Naive Bayes, decision tree, random forest, and logistic
regression, run in parallel to predict the class simultaneously. The most suitable class
is selected using a voting classifier. The model allows us to take advantage of all four
classifiers and prevents the system from overfitting.

5.5 PERFORMANCE ANALYSIS

To test the efficiency of the proposed RF-based model, precision, recall, and F1
scores of ML algorithms have been computed. Confusion matrix (CM) is created
to compute the considered performance metrics. The CM for the proposed RF
model has been depicted in Figure 5.3. It can be observed from the figure that the
proposed RF model accurately classifies more than 90% of patients which shows
its efficacy. Based on CM, precision, recall, and F-measure are also computed.
The details of the performance metrics have been presented in Table 5.1. It can
pertinent from the table that the proposed model returns better results on each
performance metric.

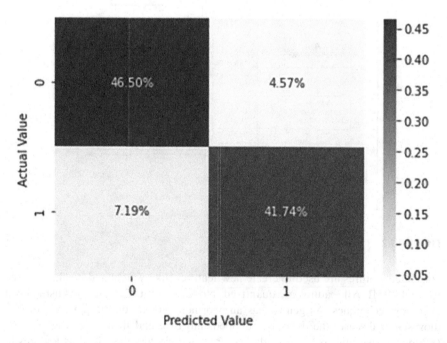

FIGURE 5.3   Confusion matrix.

TABLE 5.1
Performance Metrics

Measure	Value
Sensitivity	0.9222
Specificity	0.93
Precision	0.83
Negative predictive value	0.0778
False positive rate	0.07
False discovery rate	0.1545
False negative rate	0.88
F1 score	0.8857
Matthews correlation coefficient	0.7638

5.6 APPLICATION

Early diagnosis of chronic diseases is one of the applications of health monitoring utilizing machine learning, which can help physicians, nurses, patients, and other caregivers make better decisions. Similar to how doctors use this system, regular people can use it to evaluate whether they have serious health issues and, if so, seek care at nearby hospitals. The suggested IoT-based intelligent health monitoring system is beneficial, particularly for people living alone or elderly patients who require regular monitoring.

5.7 CONCLUSION

This chapter proposes a method to prevent the disease with early intervention rather than treatment after diagnosis. It is possible to anticipate diseases more accurately using the proposed system. This proposed system will help medical professionals decide more wisely, spot trends and developments, and boost the effectiveness of studies and clinical trials. It enhances the way healthcare services are delivered, reduces expenses, and handles patient data with extreme precision.

REFERENCES

1. Novartis, "Arogya Parivar-Improving access to healthcare." https://www.novartis.com/in-en/about/arogya-parivar (Accessed on May 5, 2022).
2. Li, J., Y. Meng, L. Ma, S. Du, H. Zhu, Q. Pei, and X. Shen. "A federated learning based privacy-preserving smart healthcare system." *IEEE Transactions on Industrial Informatics* 18, 2021, 2022.
3. Bhanuteja, G., Kaustubh Anand Kandi, K. Srinidhi, R. Dikshitha, and Anil Kumar. "CAMISA: An AI solution for COVID-19." In 2021 International Conference on Design Innovations for 3Cs Compute Communicate Control (ICDI3C), pp. 216–222. IEEE, 2021.

4. Ahmad, Muhammad, Saima Sadiq, Ala Saleh Alluhaidan, Muhammad Umer, Saleem Ullah, and Michele Nappi. "Industry 4.0 technologies and their applications in fighting COVID-19 pandemic using deep learning techniques." *Computers in Biology and Medicine* 145, 105418, 2022.

5. Palwe, Sushila, and Sumedha Sirsikar. "Industry 4.0 Technologies and Their Applications in Fighting COVID-19." In *Sustainability Measures for COVID-19 Pandemic*, pp. 237–251. Springer, Singapore, 2021.

6. Habib, Al-Zadid Sultan Bin, and Tanpia Tasnim. "An ensemble hard voting model for cardiovascular disease prediction." In 2020 2nd International Conference on Sustainable Technologies for Industry 4.0 (STI), pp. 1–6. IEEE, 2020.

7. Chern, Ching-Chin, Weng-U. Lei, Kwei-Long Huang, and Shu-Yi Chen. "A decision tree classifier for credit assessment problems in big data environments." *Information Systems and e-Business Management* 19, 363–386, 2021.

8. Belgiu, Mariana, and Lucian Drăguţ. "Random forest in remote sensing: A review of applications and future directions." *ISPRS Journal of Photogrammetry and Remote Sensing* 114, 24–31, 2016.

9. Menard, Scott. *Applied Logistic Regression Analysis*. No. 106. SAGE, New York, 2002.

10. Ontivero-Ortega, Marlis, Agustin Lage-Castellanos, Giancarlo Valente, Rainer Goebel, and Mitchell Valdes-Sosa. "Fast Gaussian Naïve Bayes for searchlight classification analysis." *Neuroimage* 163, 471–479, 2017.

11. Dong, Xibin, Zhiwen Yu, Wenming Cao, Yifan Shi, and Qianli Ma. "A survey on ensemble learning." *Frontiers of Computer Science* 14, 241–258, 2020.

12. Breiman, Leo. "Bagging predictors." *Machine Learning* 24 (2), 123–140, 1996.

6 Assessment of Fuzzy Logic Assessed Recommender System A Critical Critique

S. Gopal Krishna Patro
GIET University
Koneru Lakshmaiah Education Foundation

Brojo Kishore Mishra
GIET University

Sanjaya Kumar Panda
NIT Warangal

CONTENTS

6.1 INTRODUCTION

Recommendation systems are referred as information filtering which exhibits the social elements such as users or groups and the products such as movies, music, books based on the user interest [1]. It is otherwise stated as based on the user's recent behavior such as watching a movie, purchasing a book, or listening to music,

DOI: 10.1201/9781003321149-6

the recommender system (RS) or recommendation system works as an information filter and displayed to the user [2–5]. Various researchers have developed and utilized effective algorithms for this kind of information filtering process. The recommendation systems are used by many websites such as IMDB, Google, YouTube, and Amazon. For example, in a movies website, if a user registered and watched adventure movies regularly, then the recommendation system identifies this trend and recommends adventure movies that he had not watched so far [6,7]. For a company also, it provides a lot of advantages, and sales are boosted simultaneously. Similar things are thus found by RS related with user's likes. In today's modern world, the main problem is the information overloading, which leads to navigation problems in the ocean of all available real choices. RS has been originated as a most generous big data application that enables the user to detect the exact information in this data overflowing period. As the e-commerce business is expanded, consumers may be subjected to a huge amount of data before they make a decision for buying the items [8]. The system provides a resolution for this data overloading issue thereby suggesting better products to the consumers. The system attempts to improve the e-commerce business because it assists in the detection of Internet of Things (IoT) for the customers they want to purchase but they could not view or detect them among the overloaded data [9]. RS is generally utilized for enhancing the loyalty and trust of the customer as it guarantees exclusive personalized service. The interest of the user may be estimated by directly buying the product based on user preferences or by observing the past buying behavior [10].

In particular, two other tasks that are linked to the RS are the prediction task and the recommendation task. With such objectives, various recommendation techniques were developed on the basis of their working principles and categorized into various kinds.

The contribution toward this paper is listed as follows:

- To review various recommendation system techniques and associated performance metrics.
- To elaborate the recommendation system perspectives based on the 8 methods such as content-based, collaborative filtering, hybrid RS, social, fuzzy, genetic, group RS, context-aware, and deep learning.
- To focus on the fuzzy-based recommendation system and analyze the existing studies accordingly.
- To evaluate the recommendation approaches used in existing studies based on performance metrics.

6.2   RECOMMENDATION SYSTEM APPROACHES

The formal definition of the RS is as follows:

Consider A and B as the set of items. A function X has been defined, which quantifies the usage of the item A ($\in A$) to the corresponding user B ($\in B$) through mapping

$$X : A \ X \ B \rightarrow R_a.$$

where R_a is referred to as ratings set. The aim is to utilize and learn X to determine the previously not viewed item a by the user b for the generation of most possible group

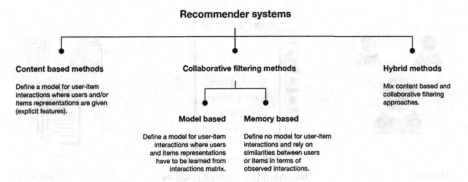

FIGURE 6.1 Algorithms: recommendation system.

of recommended items. Various conventional recommendation methods comprised collaborative filtering (CF), content-based RS (CBS), and hybrid RS (HRS). Every method possesses its own merits and demerits. The CF (which is naturally domain free) has prevented trending works from CBS. But CF has been found to be affected by sparsity and cold-start problem (CSP), whereas CBD is found to be affected by overspecialization. Several enhanced methods were suggested for the mitigation of the disadvantages such as interactive, social, deep learning fuzzy logic, and aware-based RS, which improve the accuracy and coverage. The recommendation system overview is shown in **Figure 6.**1.

The recommendation system purpose is to suggest relevant items to users. To obtain the above purpose, the major approaches followed by RC are content based and CF. The brief explanation of these two approaches is as follows.

6.2.1 Collaborative Filtering Method

For RS, the collaborative method is depending on the previously recorded interaction between the item and user with the intention to generate new recommendations. In the "user–item interactions matrix" these interactions are stored. The major idea behind the collaborative methods is the detection of similar users/items by the sufficient past user–item interactions shown in **Figure 6.**2. Further, the predictions are made based on the proximities estimated. The class of the CF algorithms is categorized into two substrategies named as memory-based and model-based approaches. The approach under model based refers as generative model, which describes the user–item interactions, and new predictions are made. The memory-based approaches are works openly with recorded interactions values, not any model assumes and depends on nearest neighbor search. For example, from user of interest, the closest user is identified, and among these neighbors, the most famous items are suggested [11].

6.2.1.1 Cold-Start Problem

In recommendation system, the CSP is the common problem. This problem occurs when the new items or new users arrive in e-commerce platforms. The CF considers

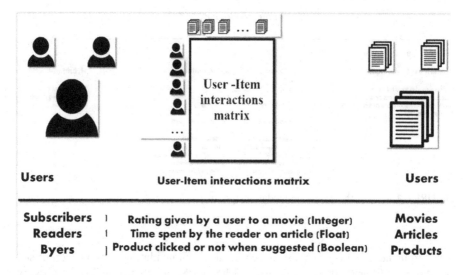

FIGURE 6.2   User/items interactions matrix.

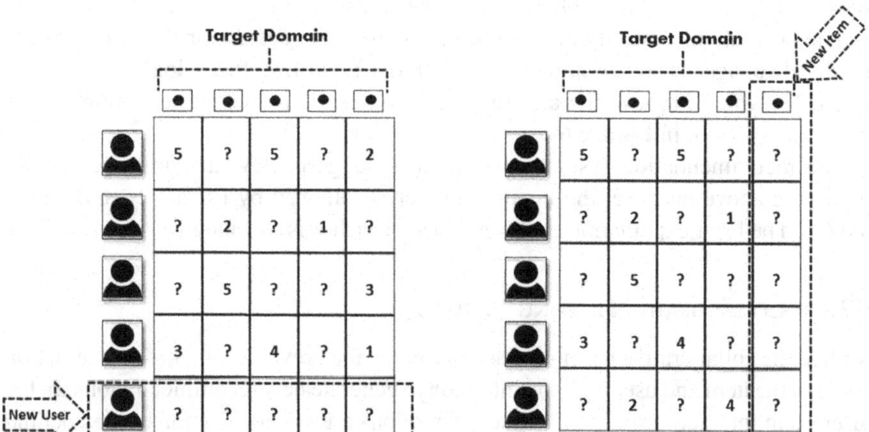

FIGURE 6.3   Representation of cold-start problem in RS.

each item or each user had some ratings so that similar user ratings are derived even if the ratings are unavailable. It happens when the system does not know the data about new user with respect to new items [12].

Figure 6.3, represents the CSP, and it happens when new user enters the system without preference data for that user. In this figure, rows represent the customers or users and columns represents the products. It had new user problem and new item problem. CF provides the solution for this CSP [13]. It predicts the ratings to target user. CF extracts the patterns of similarities and relationships among users and utilized learn predictive model to generate recommendations. Some predictions are calculated for pair of user items and for every unknown rating by using rating matrix. The collaborating system filters the items based on highly predicted ratings and it recommends to user. The CSP is also called as new user problem. The data sparsity is considered to rectify the issues of cold start. The deep learning and CF are used for CSP [14]. CF is suffered from "CSP" as for recommendations, it considers past interactions. To the new users, any kind of recommendation is impossible or to any user recommending a new item and few interactions corresponding to users or items to be handled efficiently. This limitation is handled by random strategy – random items recommendation to new users or to random users with new items, maximum expectation strategy – recommending the most popular items to new users or to most active users with new items, exploration strategy – to new users recommending a set of different items or to set of different users with new item. For the early user life or the item, the noncollaborative method is used.

6.2.2 CONTENT-BASED METHODS

The content-based approaches utilized the extra information about users/items like, for example, in movies RS, this extra information is any other personal data such as age, sex, and other information about users and also the main actors, movies description, or the duration, whereas the collaborative methods are based only on the interactions between user and item. Based on the available features, the content-based methods build a model that describes about witnessed user–item interactions, as shown in **Figure 6.**4.

The content-based approach has been based on the description of the item and user preference of the profile. For generating recommendation, potential items have been compared with the rating items for determining the matching of the item. In this method, the user interest is oriented on the features based on objects rated. Lee and Lee presented OTB (outside-the-box) recommendations through the generated regions of items with respect to the items similarity and the user adherence to the

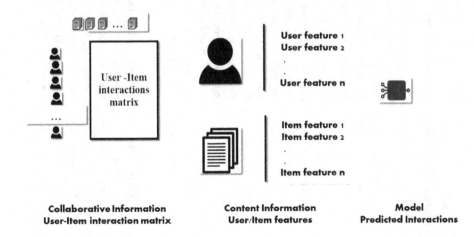

FIGURE 6.4   Content-based approach.

TABLE 6.1

Comparative Analysis of Existing Recommendation Studies

Reference	Technique	Description	Benefits	Limitation
[17]	Ontology-dependent recommendation	Ontology takes different structures in light of the unique situation. Utilizing the Web Ontology language contrasted and different dialects the philosophy addressing the XML plot in data set setting.	Enhance quality of personalized recommendation. Time-consuming	Need for domain knowledge. Distinguishing the right suggestion approach mix and assessing the presentation is troublesome in the event of hybrid recommendation-based ontology.
[18]	Context-aware recommendation	This system comprised of recommendation provider for gathering the contextual data and feedback data from sensor and learner application module.	Better performance in hybrid recommender system. Recommendations can be adjusted based on context	Integrate contextual information
[19]	Hybrid filtering	This study zeroed in on book proposal frameworks in light of crossover approach contained cooperative and content-based expectations and compelling mix came about.	No cold-start issue is seen. Combined the Spark big data platform and obtains the personalized book recommendation and further utilization rate is enhanced.	Time complexity issue is seen.
[20]	Demographic recommendation	This study zeroed in on monetary preparation and for that proposal cycle needs to change and worked on in better manner.	It does not need rating history of particular learner. For various online and offline applications, this system has used.	Learner must retrieve personal attributes.
[21]	Collaborative filtering recommendation	Different algorithmic techniques featured which is used to upgrade data recovery to give student a proficient suggestion by fulfillment level and execution improvement.	Domain knowledge is not required. Effectiveness obtained by improved information retrieval.	Effective utilization of recommender system is not seen in case of comparisons with other domains. Over specification
[22]	Fuzzy-based recommendation system	A legitimate watchword extraction method is created in this review for suggesting the items in a sure and negative way to clients.	Efficient classification accuracy resulted.	Multimodal data is yet to be considered to improve the similar products search. In an organizational setting practical application required

regions that overwhelms the chances in the framework [15]. This stated chance from the returning of the items which are more similar to the existing user rates. This method possesses a very serious drawback of minimized scope. Due to that, it may recommend the similar items with the original seed in accordance with the user altering preference and apart from that it does not recommend the very popular products to the users. Further the naive user issue could not be suggested unless adequate items were rated by the user for understanding the user preference [16] (Table 6.1).

6.3 VARIOUS RECOMMENDATION SYSTEM BASED ON E-COMMERCE

The following sections deal with different RSs with e-commerce:

6.3.1 PERSONALIZED RECOMMENDATION SYSTEM OF E-COMMERCE BASED ON BIG DATA ANALYSIS

Here, the analysis was segmented into four levels such as presentation layer, business layer, management layer, and the data layer. All these layers are related closely to big data. Related to the accuracy of the electronic commerce system, the e-commerce framework depending on big data was built 5.

Personalized recommendation algorithm (PRA) for e-commerce: personalized recommendation algorithm for commodity n and keyword t in e-commerce is as follows:

$$|v_{minimum}| \leq 1 \tag{6.1}$$

$$|q_{tm}| \leq 1 \quad , \text{where } m = 1,2,3,\ldots,M$$

$$|c_{nt}| \leq 1 \tag{6.2}$$

Here, c_{nt} represents the similarity of goods. If n and t were similar in M characteristics, then membership of every feature is the same "−" or same "+", and the corresponding product is "+".

$$vc_{nt} >= \text{MAXIMUM}() * \text{MAXIMUM}(q_{tm}) \tag{6.3}$$

Generally, the c_{nt} is larger. Suppose if n and t are not same, then it represents that they are alternate in some characteristics, and membership of them was an alternate sign, then the corresponding product was "−", then n_{tc} was smaller. When membership n and keyword t equal to 1 only in characteristic, where other was 0, then $c_{nt} = 1$, and n and t were entirely the same; where membership of n and t is 1 and −1 in characteristic, other was 0, then $c_{nt} = -1$, and n and t were not entirely same. So, for keyword t, it could recommend suitable PRA by choosing larger items of c_{nt}. This technique provides a good result for a small range but with a variety of goods the technique address complexity to grasp while choosing computing membership and representative features.

6.3.2 Hybrid Ontology-Dependent RS for E-Commerce

In this study [23], constructs RS with a combination of the K-nearest neighbor (KNN) algorithm and ontology. Hybrid ontology-dependent RS begins with user profile creation containing attributes, types, and items purchased by the user [24]. To suggest items to the user, there was a necessity to predict other users who had purchased at least 1 common item and chose the one indicating past criteria who have also purchased other items. These were items suggested to the user [25]. Once creating the user and his neighbor profile, the KNN algorithm was used to predict the user's nearest neighbor. This procedure facilitates extracting items that were not purchased by the user. Finally, KNN was utilized to predict the nearest item related to the user's interest, which contains the overall rating provided by that particular neighbor, price of items on those items. Additionally, a text review of every product was included since it acts as a large source to extract needful information [26].

6.3.3 Trust-Dependent CF Algorithm for E-Commerce RS

RS has attained success in resolving the issue of information overload, but still, there were a few problems such as cold start and data sparseness. In the case of a sparse rating, it is difficult to receive a satisfying result in RS [27]. One way to resolve this issue was to present trust within RS. To compute user similarities, Pearson correlation coefficient was utilized to state trust metrics. Available trust metrics depend on the general guess that information given by the user was accurate, true, and able to reflect the real interest of the user. However, in numerous cases, this guess was not reasonable. So, in a way to model good standing measure, more data about rating and user must be considered. This existing study takes rating data's reliability and introduced trust-dependent design depending on the CF algorithm.

Currently, there were numerous fake ratings on e-commerce websites [28]. Here, the fake rating was classified as follows:

- Because of on-sale activity, users would receive some cashback if they provide high ratings to items.
- Purposely hiring someone to rate products.

  For resolving second type of fake rating, trust-dependent recommendation design with CF takes the following features such as first, the design integrates relationship among trusted users, and degree of trust for purpose of rating. User similarity was a trust relationship among users. On the other hand, the degree of trust for rating was stated from two features: one was to spot fraudulent users and eliminate their rating. Second was to give metric for every rating's trust-dependent strength depending on votes given by other users. Later, enhanced slope one algorithm was presented depending on the trust-dependent model.

6.3.4 ENHANCED SLOPE ONE ALGORITHM

Measures user similarity by recording the rating matrix of the user's item and then measures similarity. In the rating matrix of the user's item, items rated by users are represented as user vector, then every user was indicated as n dimension rating vector, that is $V_q = \left(s_{q1},\ s_{q2},\dots, s_{qn} \right)$, s_{qn} is rating user V_q for item J_n. Then, user similarity was measured depending on the user–item rating matrix. This kind of algorithm was utilized in various applications like RS for the social network [27,29–31], or location-dependent service [32].

6.3.5 SIMILARITY MEASURES

Similarity measures were utilized to predict the likeliness of two users. Generally, similarity measures are utilized only when users have rated both items considered. This creates a bottleneck for new users.

In CF, detection for user j in the nonpersonalized and personalized state is given as follows:

$$Q\ (b,\ j) = \sum_{v=1}^{m} \frac{s(v,j)}{m} \tag{6.4}$$

Here Q is a detection for user b for the item j and detection is made for user b depending on other users m, $s(v, j)$ is a rating by user v in item j.

Now, there are two considerations, first, if j is not liked by everyone and has various opinions about j; second, if the rating utilized for j is not on the same scale. These cases must be managed differently.

In first case, for detection j to b, if the item was not drawn equally from entire users, then personalized weight was utilized for each user and finally, it is added. This is expressed as follows:

$$Q\ (b,\ j) = \frac{\sum_{v=1}^{m} s(v,j) * o(b,\ v)}{\sum_{v=1}^{m} o(b,v)} \tag{6.5}$$

Here, $s(v, j)$ is rating done by v for j and $o(b, v)$ is weighted arrangement among users b and v. It is provided by similarity among users b and v.

In second case, where rating is not on the same scale, the difference of rating of user v for j item and the average rating for user v is to consider as follows:

$$Q(b,\ j) = \sum_{v=1}^{m} s(v,j) - \frac{s(v)}{m} + s(b) \tag{6.6}$$

Here, (v, j) denotes user–item pair, and $s(b)$ provides deviation which was added to the average rating of user b.

CF states that user's previous agreements are utilized to detect their further agreements. Every user remains stable and cannot modify over time and is also synchronized with one another. So, in CF, the system must work carefully in the area of agreement. Despite this, there are also cases, where people agree on one domain and not on other domains [33]. For example, few peoples feel good to discuss politics but are not interested to have technical discussions. In such a situation, detection for j was provided by equations 6.2 and 6.3.

6.4 CONCLUSIONS

The evolution of websites results in the data overloading and hence RS offers a comprehensive solution for filtering the required information and promotes its usage. In this chapter, the recommendation system approaches, similarity measures, evaluations, and substantially, the fuzzy logic-based recommendation system are analyzed and reviewed. The study also discussed the merits and demerits of several RS techniques to offer a panorama by which the researchers, educators, practitioners, and industrialists could able to quickly understand and learn the scheme behind the RS to ease the processing. This chapter is then attempted to discuss every key aspect of the recommendation system for its improvement. This current study is considered as starting point for the improvement of more recent contributions in the emerging field of the recommended systems supported by fuzzy tools.

REFERENCES

1 H. Y. Jeong, K. M. Park, M. J. Lee, D. H. Yang, S. H. Kim, and S. Y. Lee, "Vitamin D and hypertension," *Electrolyte Blood Press.*, vol. 15, no. 1, pp. 1–11, 2017, doi: 10.5049/EBP.2017.15.1.1

2. M. Scholz, V. Dorner, G. Schryen, and A. Benlian, "A configuration-based recommender system for supporting e-commerce decisions," *Eur. J. Oper. Res.*, vol. 259, no. 1, pp. 205–215, 2017, doi: 10.1016/j.ejor.2016.09.057

3. Y. Guo, Y. Liu, A. Oerlemans, S. Lao, S. Wu, and M. S. Lew, "Deep learning for visual understanding: A review," *Neurocomputing*, vol. 187, pp. 27–48, 2016, doi: 10.1016/j.neucom.2015.09.116

4. C. J. Carmona, S. Ramírez-Gallego, F. Torres, E. Bernal, M. J. Del Jesus, and S. García, "Web usage mining to improve the design of an e-commerce website: OrOliveSur.com," *Expert Syst. Appl.*, vol. 39, no. 12, pp. 11243–11249, 2012, doi: 10.1016/j.eswa.2012.03.046

5. T. K. Paradarami, N. D. Bastian, and J. L. Wightman, "A hybrid recommender system using artificial neural networks," *Expert Syst. Appl.*, vol. 83, pp. 300–313, 2017, doi: 10.1016/j.eswa.2017.04.046

6. J. Lv, B. Song, J. Guo, X. Du, and M. Guizani, "Interest-related item similarity model based on multimodal data for top-N recommendation," *IEEE Access*, vol. 7, no. c, pp. 12809–12821, 2019, doi: 10.1109/ACCESS.2019.2893355

7. X. Li and J. She, "Collaborative variational autoencoder for recommender systems," *Proc. ACM SIGKDD Int. Conf. Knowl. Discov. Data Min.*, vol. Part F1296, pp. 305–314, 2017, doi: 10.1145/3097983.3098077

8. G. Shani, D. Heckerman, and R. I. Brafman, "An MDP-based recommender system," *J. Mach. Learn. Res.*, vol. 6, pp. 1265–1295, 2005.

9. G. van Capelleveen, C. Amrit, D. M. Yazan, and H. Zijm, "The recommender canvas: A model for developing and documenting recommender system design," *Expert Syst. Appl.*, vol. 129, pp. 97–117, 2019, doi: 10.1016/j.eswa.2019.04.001

10. S. Parvatikar and D. Parasar, "Recommendation system using machine learning," *Int. J. sArtif. Intell. Mach. Learn.*, vol. 1, no. 1, p. 24, 2021, doi: 10.51483/ijaiml.1.1. 2021.24-30

11. J. Guo, J. Deng, and Y. Wang, "An intuitionistic fuzzy set based hybrid similarity model for recommender system," *Expert Syst. Appl.*, vol. 135, pp. 153–163, 2019, doi: 10.1016/j. eswa.2019.06.008

12. K. Xinchang, P. Vilakone, and D. S. Park, "Movie recommendation algorithm using social network analysis to alleviate cold-start problem," *J. Inf. Process. Syst.*, vol. 15, no. 3, pp. 1–16, 2019, doi: 10.3745/JIPS.04.0121

13. A. K. Pandey and D. S. Rajpoot, "Resolving cold start problem in recommendation system using demographic approach," *2016 Int. Conf. Signal Process. Commun. ICSC 2016*, pp. 213–218, 2016, doi: 10.1109/ICSPCom.2016.7980578

14. J. Wei, J. He, K. Chen, Y. Zhou, and Z. Tang, "Collaborative filtering and deep learning based recommendation system for cold start items," *Expert Syst. Appl.*, vol. 69, pp. 1339–1351, 2017, doi: 10.1016/j.eswa.2016.09.040

15. K. Lee and K. Lee, "Escaping your comfort zone: A graph-based recommender system for finding novel recommendations among relevant items," *Expert Syst. Appl.*, vol. 42, no. 10, pp. 4851–4858, 2015, doi: 10.1016/j.eswa.2014.07.024

16. J. Son and S. B. Kim, "Content-based filtering for recommendation systems using multiattribute networks," *Expert Syst. Appl.*, vol. 89, pp. 404–412, 2017, doi: 10.1016/j. eswa.2017.08.008

17. G. George and A. M. Lal, "Review of ontology-based recommender systems in e-learning," *Comput. Educ.*, vol. 142, p. 103642, 2019, doi: 10.1016/j.compedu.2019. 103642

18. Y. Chen, X. Li, J. Liu, and Z. Ying, "Recommendation system for adaptive learning," *Appl. Psychol. Meas.*, vol. 42, no. 1, pp. 24–41, 2018, doi: 10.1177/0146621617697959

19. Y. Tian, B. Zheng, Y. Wang, Y. Zhang, and Q. Wu, "College library personalized recommendation system based on hybrid recommendation algorithm," *Procedia CIRP*, vol. 83, pp. 490–494, 2019, doi: 10.1016/j.procir.2019.04.126

20. X. Li, M. Gao, W. Rong, Q. Xiong, and J. Wen, "Shilling attacks analysis in collaborative filtering based web service recommendation systems," *Proc. -2016 IEEE Int. Conf. Web Serv. ICWS 2016*, pp. 538–545, 2016, doi: 10.1109/ICWS.2016.75

21. H. F. Nweke, Y. W. Teh, M. A. Al-garadi, and U. R. Alo, "Deep learning algorithms for human activity recognition using mobile and wearable sensor networks: State of the art and research challenges," *Expert Syst. Appl.*, vol. 105, pp. 233–261, 2018, doi: 10.1016/j. eswa.2018.03.056

22. L. Kolhe, A. K. Jetawat, and V. Khairnar, "Robust product recommendation system using modified grey wolf optimizer and quantum inspired possibilistic fuzzy C-means," *Cluster Comput.*, vol. 24, no. 2, pp. 953–968, 2021, doi: 10.1007/s10586-020-03171-6

23. M. Guia, R. R. Silva, and J. Bernardino, "A hybrid ontology-based recommendation system in e-commerce," *Algorithms*, vol. 12, no. 11, pp. 1–19, 2019, doi: 10.3390/ a12110239

24. I. Huitzil, F. Alegre, and F. Bobillo, "GimmeHop: A recommender system for mobile devices using ontology reasoners and fuzzy logic," *Fuzzy Sets Syst.*, vol. 401, pp. 55–77, 2020, doi: 10.1016/j.fss.2019.12.001

25. M. Nilashi, O. Ibrahim, and K. Bagherifard, "A recommender system based on collaborative filtering using ontology and dimensionality reduction techniques," *Expert Syst. Appl.*, vol. 92, pp. 507–520, 2018, doi: 10.1016/j.eswa.2017.09.058

26. G. D. Congying Guan, Shengfeng Qin, Wessie Ling, "Apparel recommendation system evolution : An empirical review Article information : About Emerald www. emeraldinsight.com," *Int. J. Cloth. Sci. Technol.*, 2016. doi: 10.1108/ijcst-09-2015-0100

27. L. Jiang, Y. Cheng, L. Yang, J. Li, H. Yan, and X. Wang, "A trust-based collaborative filtering algorithm for E-commerce recommendation system," *J. Ambient Intell. Humaniz. Comput.*, vol. 10, no. 8, pp. 3023–3034, 2019, doi: 10.1007/s12652-018-0928-7

28. R. Katarya and O. P. Verma, "An effective web page recommender system with fuzzy c-mean clustering," *Multimed. Tools Appl.*, vol. 76, no. 20, pp. 21481–21496, 2017, doi: 10.1007/s11042-016-4078-7

29. Y. Zhang, H. Abbas, and Y. Sun, "Smart e-commerce integration with recommender systems," *Electron. Mark.*, vol. 29, no. 2, pp. 219–220, 2019, doi: 10.1007/s12525-019-00346-x

30. J. Cai, Y. Wang, Y. Liu, J. Z. Luo, W. Wei, and X. Xu, "Enhancing network capacity by weakening community structure in scale-free network," *Futur. Gener. Comput. Syst.*, vol. 87, pp. 765–771, 2018, doi: 10.1016/j.future.2017.08.014

31. S. Peng, A. Yang, L. Cao, S. Yu, and D. Xie, "Social influence modeling using information theory in mobile social networks," *Inf. Sci. (Ny).*, vol. 379, pp. 146–159, 2017, doi: 10.1016/j.ins.2016.08.023

32. T. Peng, Q. Liu, D. Meng, and G. Wang, "Collaborative trajectory privacy preserving scheme in location-based services," *Inf. Sci. (Ny).*, vol. 387, pp. 165–179, 2017, doi: 10.1016/j.ins.2016.08.010

33. J. Iwanaga, N. Nishimura, N. Sukegawa, and Y. Takano, "Improving collaborative filtering recommendations by estimating user preferences from clickstream data," *Electron. Commer. Res. Appl.*, vol. 37, p. 100877, 2019, doi: 10.1016/j.elerap.2019.100877

7 Intelligent Analytics in Big Data and Cloud
Big Data; Analytics; Cloud

K. Aditya Shastry and B. A. Manjunatha
Nitte Meenakshi Institute of Technology

CONTENTS

DOI: 10.1201/9781003321149-7

7.1   INTRODUCTION

The volume of information produced, processed, and exchanged has increased since the dawn of the digital revolution. Large amounts of information can be found in a variety of places, including spreadsheets, sites, and forums, including acoustic feeds. As a consequence of increased data collection, vast volumes of widespread and complicated data are generated, that must be effectively generated, saved, distributed, and evaluated in order to mine meaningful knowledge. Such information has enormous promise, but it is becoming increasingly sophisticated, insecure, and risky, as well as becoming irrelevant. Given that this research may encompass accessibility and study of clinical records, engagements, financial documents, and public records, including genomic patterns, the advantages, and constraints of accessing this information are debatable. The notion of big data processing and analytics was born out of the need for efficient and effective analytics services, apps, development environments, and platforms [1].

Big data analytics (BDA) is being used in a variety of sectors and areas. Health research, transportation systems, and worldwide safety, as well as the modeling of challenges in the cultural and ecological sectors, are just a few of the uses. Aside from standard applications in industry, commerce, and societal management, current research is focusing on the potential uses of "big data" in practice [2]. Among them most important potential uses of "big data analytics" and "cloud computing" is in the biological sciences [3]. Among the high-impact fields defined are tissue engineering, structural and posttranslational modification predictions, tailored therapy, and meta-transcriptomics. Aside from that, among the major uses of "big data analytics" is to enhance the client satisfaction levels of prevailing business models.

By design, "big data" refers to a variety of information types, including formatted, quasi, and unstructured information, resulting in a complicated data architecture [4]. Such infrastructure's complexities necessitate sophisticated administration and technological solutions. The multi-V paradigm forms one of the most widely used frameworks for interpreting huge datasets. The multi-V model is depicted in Figure 7.1.

"Variety", "volume", "velocity", "veracity", "value", "viability", "validity", and "volatility" are the Vs utilized to describe "big data" [5]. "Variety" is determined by the many types of information present in a database, whereas "velocity" is determined by the frequency with which information is produced. The quantity of

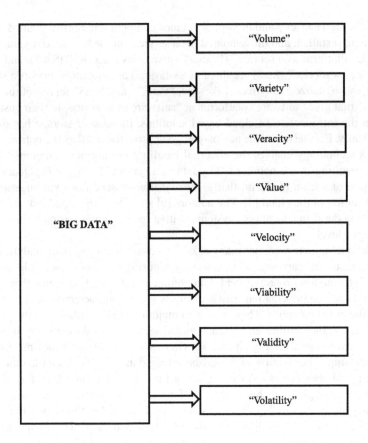

FIGURE 7.1 Six V's of big data.

information is referred to as volume. Veracity and value are two more attributes that reflect accuracy of information and usefulness in relation to big data mining, correspondingly. "Viability" is the ability of the model to signify reality. "Validity" refers to the accuracy and correctness of the data. "Volatility" refers to the time duration for which the data is useful or relevant. Furthermore, the researchers [6] provided additional characterization known as the "HACE" theorem. Big data includes two key properties, as per this hypothesis. To begin with, it has a significant quantity of complex information from various diverse sources. Secondly, the information is decentralized and dispersed.

On the Internet, with all the apps created on it, information is the most important component of cooperation and interaction. The enormous growing importance of information-centric services like "Facebook", "LinkedIn", "Twitter", "Amazon", "eBay", and "Google+" is contributing to an increase in the demand for data processing and storage on the cloud. Schouten [7] cites Gartner's prediction that by 2016, one-half of all information will be stored in the "cloud".

Furthermore, the big data analytics data mining methods have powerful computational demands. As a result, high-performance CPUs are required to complete the task. The cloud is an excellent platform for storing, analyzing, and interpreting large amounts of information, since it meets two of the most important needs of "big data analytics": high capacity and high-performance computation. Software and data apps are written, installed, and implemented "as a service" in the "cloud computing" environment. "Platform as a service" (PaaS), "software as a service" (SaaS), and "infrastructure as a service" (IaaS) are three cross-layered architectures. IaaS is a strategy for renting virtualized resources. Likewise, "PaaS" and "SaaS" represent the "cloud services" that give "software" platforms or "software as a service" to their customers.

With the introduction of cloud-based solutions, the cost of storage has dropped dramatically. Furthermore, the pay-as-you-go approach as well as the notion of commodities technology enables the effectual handling of enormous amounts of information, resulting in the notion of "big data as a service". "Google BigQuery" is an illustration of such a platform, that gives real-time observations via huge amounts of information in the cloud [8]. The authors [9] show how the cloud may be used to handle "big data" in educational systems, with a specific emphasis on information at the campus level.

Nonetheless, there were not many effective "cloud-based big data analytics" apps yet. As a result, researchers are increasingly focusing on cloud-based "big data analytics". Information protection and data confidentiality are two issues that are visible in this setup. Information trustworthiness is also characterized as a service as part of the cloud solutions. There will be a major drop in credibility as the potential for security vulnerabilities and privacy violations increases dramatically if big data methods are implemented in the cloud. There will be a major problem of control and ownership. The promise of "cloud-based big data analytics", on the other hand, has prompted researchers to examine prevailing difficulties and look for solutions. The numerous angles and characteristics of data mining methodologies' usage in the cloud infrastructure for "big data analytics" are discussed in this paper. In addition, it investigates ongoing studies and vulnerabilities associated with them, including study instructions prospects in "cloud-based big data analytics".

Due to the high general amount and intricacy of the information involved, conventional database technologies, including computation or statistical methods, could not be employed for "big data analytics". Classic business intelligence solutions involve "OLAP", "BPM", "mining", and repository systems, including "Relational Data Base Management System (RDBMS)", that are associated with traditional statistical approaches and methods.

"MapReduce" is one of the most prominent models for information processing on a network of machines. An overview of modeling techniques that facilitate large data analytics is provided [10]. Although "MapReduce/Hadoop" is identified as one of the most productive models for "big data analytics", it also adds other dialects and modules such as "HiveQL", "Latin", and "Pig" that suggest substantial improvements for this task. "Hadoop" is merely a transparent version of the "MapReduce" context, which was designed as a shared file system in the first place. As per the work [11], "Hadoop" has progressed into a whole environment or architecture that operates with "MapReduce" components and comprises a variety of technological tools such as the "Hive" and "Pig" languages, a synchronization service known as "Zookeeper", and a networked database store known as HBase. "Google MapReduce", "Spark", "Hadoop", "Twister", "Hadoop Reduce", and "Hadoop++" are certain frameworks available for cloud-based big data analytics [12]. The application of cloud technology in "big data analytics" is illustrated in Figure 7.2.

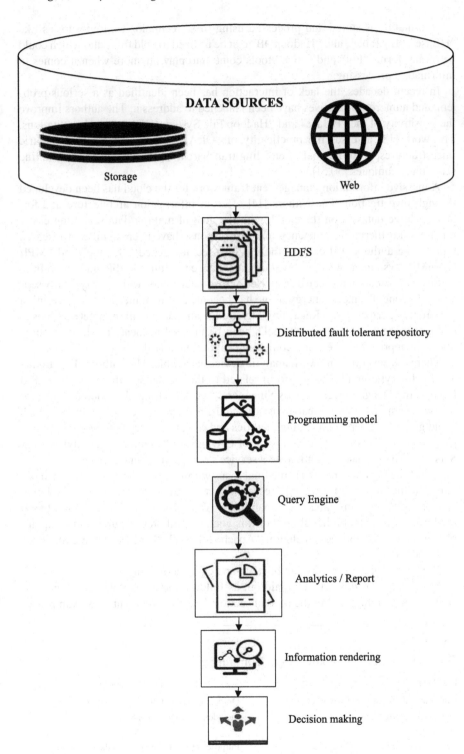

FIGURE 7.2   Application of cloud technology in big data analytics.

Information is stored and processed using these technologies. Repositories like "HBase", "BigTable", and "HadoopDB" can be utilized to hold this data, which could be of any form. "Pig" and "Hive" tools come into play anymore when it comes to information processing.

In recent decades, the lack of interaction has been identified as a serious problem, and numerous initiatives have been launched to address it. The authors improve the sensitivity of the "HBase" and "Hadoop File System (HDFS)" implementations. The work [13] assesses the practicality of "OLAP Web" Apps for cloud-centric infrastructures, with the goal of enabling transparent and widespread use of online scientific techniques [50,51].

A massive information management framework for the cloud has been developed through investigation. The authors [14] offer an information architecture and format for large datasets on the cloud, with the goal of making data searching easier for users. Furthermore, efficiency and fast response have been significant research topics. The authors [15] explore the use of a recommended "Hadoop" and "MPI/OpenMP" system, as well as how it might enhance the functionality and reliability.

Since information must be transported across datacenters, which are typically separated by long distances, energy usage has become a critical metric for determining the process performance. Regarding big storage and computation, a network-based navigation method known as "GreeDi" can be utilized to locate the best resource optimum approach to the cloud computing environment [16].

There are several framework analytic devices available. The authors [17] provide a "direct acrylic graph (DAG)" form related to the technology that was utilized to forecast the Dengue spread across Singapore. "Aneka" [18] and "CloudComet" [19] are two examples of live threat analytics and the necessity for an architecture, which could give consumers with coding facilities and capital to carry out the same. In this [20], the researchers look at the notion of CAAAS, or "Continuous Analytics as A Service", which is used to anticipate a service's or a participant's behavior.

Real-time big data analysis is the final focus within "big data analysis", which has attracted the interest of academic researchers. Several corporate software solution vendors offer real-time analytic capabilities. AWS Kinesis [21] is an AWS-based platform for live stream handling. For this goal, several architectures and computer platforms were established, including "Apache S4" [22], "IBM InfoSphere Streams" [23], and "Storm" [24].

Table 7.1 lists the gap between big data and cloud computing.

The balance of this article is laid out as follows: the methodologies in big data analytics are highlighted in the second section. The final segment uses examples to

TABLE 7.1
Gap between Big Data and Cloud Computing

Big Data	Cloud Computing
Huge data increasing rapidly with time	On-demand availability of computing resources
Includes structured, unstructured, and semi-structured data	Includes IaaS, PaaS, and SaaS
Purpose is organization of large data	Purpose is to store and process data in cloud

demonstrate how big data and cloud-based intelligence could be used. The chapter's synopsis appears at the end.

7.2 CLOUD-BASED "BIG DATA" DEVICES

Three techniques for "big data analytics", "big data storage", and "big data warehouse" were chosen from the stated prominent suppliers in this part, including their succinct advantages and functions [25].

7.2.1 "AMAZON WEB SERVICES (AWS)"

AWS offers a variety of BDA frameworks that make developing and delivering BDA apps simple and fast.

7.2.2 AMAZON ELASTIC SEARCH SERVICE (BIG DATA ANALYTICS FRAMEWORK)

"Amazon ElasticSearch (Amazon ES)" is an open-source search and managed service for building domains, as well as establishing, managing, and scaling AWS cloud ElasticSearch clusters [26], with features such as continuous integration management, log analytics, and click-stream analytics. Some of the benefits include integration with other AWS services, ease of use, expandable clustering, ease of access, and compatibility with open-source APIs and applications [28–31].

7.2.3 "AMAZON S3" (BIG DATA STORAGE FRAMEWORK)

"Amazon S3" is a density and efficiency, secure, and reliable "big data storage" service with a diverse set of algorithms and applications, including big data analytics. Elastomeric leadership with more individual innovative of administration and storage business solutions, unparalleled reliability, consistency, expandability, in-place review, and assimilation with more companies can successfully, often these broad cybersecurity proficiencies, and streamlined and speeded up data transmission are just a few of the benefits of utilizing this structure [26–28,30,32,33].

7.2.4 AMAZON REDSHIFT (DATA WAREHOUSING FRAMEWORK)

"Amazon Redshift" is a well-known, highly scalable "AWS data warehouse" platform that enables rapid, simple, and cost-effective research methodology through the use of traditional SQL and popular advanced analytics. Rapid data speed, price, a straightforward and highly secure architecture, a flexible and expandable cluster, interoperability with a variety of SQL users, and huge extendibility are just a few of "Amazon Redshift's" features [26,30,34,35].

7.2.5 "AWS IDENTITY AND ACCESS MANAGEMENT (IAM)"

IAM is a web service that helps you securely control access to AWS resources. You use IAM to control who is authenticated (signed in) and authorized (has permission) to use resources. IAM provides shared access to AWS accounts, granular permissions, secure access to AWS resources for applications that run on Amazon EC2, multifactor authentication (MFA), and identity federation, to name a few [52].

7.2.6 GOOGLE CLOUD PLATFORM (GCP)

"GCP" provides a number of useful tools for a wide range of tasks, including "big data analytics", "data warehouses", "repository and archiving", and much more is a list of the most common "big data analytics", "big data storage", and "big data warehousing" framework technologies. As its name suggests, "Google Cloud Dataproc" is a completely structured and controlled cloud-centric "Apache Hadoop" and "Spark" solution for quick, simple, and cost-effective group organizational processes. Rapid ensemble scalability, cost effectiveness, and an open-source architecture are some of its characteristics and advantages [30,31,35].

7.2.7 "GOOGLE CLOUD STORAGE" (BIG DATA STORAGE FRAMEWORK)

"Google Cloud Storage" is an integrated object storage designed primarily for organizations and developers that performs a variety of activities, including actual data processing, data archiving (with "Coldline" and "Nearline" storage solutions), and analytics. Accessibility with a cheap price, streamlined preservation, as well as storage, fast and easy information transfer, pricing, improved security for corporation assets, and collaboration with top companies are just a few of the advantages of using it [30,32,33,35].

7.2.8 "GOOGLE BIGQUERY" (BIG DATA WAREHOUSING FRAMEWORK)

It is a fully organized, massively scalable, fast, and low-bandwidth repository for business intelligence in operation. Rapid architecture installation, easy expansion, and analysis with fast analytics, including business data and capital security, are just a few of the benefits of adopting this solution [30,35].

7.2.9 "IBM CLOUD"

The "IBM cloud" provides a wide range of cloud-centric "big data solutions", as well as the right instrument for the task, including the "big data analytics" framework, connectivity, and analytics. Like the paradigms chosen by cloud vendors previously explored, one paradigm for each of "big data analytics", "big data storage", and "big data warehousing" is chosen and described below.

7.2.10 "ANALYTICS ENGINE" (BIG DATA ANALYTICS FRAMEWORK)

It is one of the IBM cloud products that helps businesses with fast information processing and the resolution of various big data difficulties. It's a dedicated clustering solution that interfaces with "Apache Hadoop" and "Apache Spark" to make it easier to design and operate analytics solutions. Accessible, expandable groups and a programmable ecosystem are among its characteristics [31,36,37].

7.2.11 "IBM Cloud Object Storage" (Big Data Storage Framework)

This IBM-developed cloud-based storage platform allows users to store, manage, and access data using REST-based APIs and IBM's self-service site. This platform could link directly to apps as well as link to certain other IBM cloud services. Endurance as well as stability, geographic resiliency, memory versatility, rest-based APIs and SDKs, and IBM cloud administration interface are some of its major characteristics [33,36,37].

7.2.12 "IBM Db2 Warehouse" on Cloud (Big Data Warehouse Framework)

This is an enterprise-class cloud-centric "data warehouse" facility powered by IBM CLU Acceleration for unmatched throughput. This solution is well-organized and safe, works with "Oracle" and "Nerezza", is created for the distributed environment, serves as an information storage for data scientists, and relieves analytical capacity [36,37].

7.3 MICROSOFT AZURE

"Microsoft Azure" provides a diverse set of cloud programs that benefit designers and IT experts for building, launching, and trying to manage various applications, from mobile apps to ISC (Internet-scale computing) keys, all through its global network of datacenters and with the help of "DevOps" and unified tools. "BDA", "big data storage", and "big data warehouse" solutions are also chosen, and their characteristics and advantages are described as follows.

7.3.1 "Azure HDInsight (Big Data Analytics Solution)"

In contrast to other modern facilities, which only provide service level agreement (SLA) for essential virtual servers, our solution provides 99.99% SLA for a single example of a "virtual machine". It, for example, allows for the building of optimized groups for "Spark", "Hadoop", "Kafka", "HBase Storm", and "Microsoft R Servers", with a 99.9% SLA. Global availability, strong cybersecurity, a very efficient environment for research and innovation, pricing, and remarkable adaptability are just a few of its distinguishing characteristics [31,38,39,40].

7.3.2 "Azure Blob Storage (Big Data Storage Solution)"

"Blob Storage" is a simple and price-operative storage solution for "exabytes" of unstructured data such as audio, movies, photos, and more, organized into three tiers dependent on the regularity of network connectivity: warm, cooling, and archives. Numerous blob types, such as site, restrict, and append blob, offer memory optimization flexibility as well as fully automated spatial for simple equality of enhanced local and global connectivity [30,38,39].

TABLE 7.2
Frameworks for "Cloud-Based Big Data" Enterprise Solutions [25]

Sl. No. Architecture and Operations	"AWS"	"GCP"	"IBM Cloud"	"MS Azure"
[-5ex]1 "Big Data Analytics"	"Amazon Elastic Search Service"	"Google Cloud Dataproc"	"IBM Analytics Engine"	"Azure HDInsight"
Mode of Application	Publicly available	Publicly available	Publicly available	Publicly available
Information types	Organized, quasi, and unsupervised information	Organized, quasi, and unsupervised information	unsupervised	unsupervised
Origins of information	"Amazon S3", "Amazon Kinesis Firehose", and "Amazon DynamoDB"	"Google Bigtable", "Google Cloud Storage", and "Google Big Query"	IBM "Cloud Object Storage"	"Blob Storage"
"Operating System"	"CentOS", "Ubuntu", and "Amazon Linux"	"Debian 8"	"CentOS 7"	"Ubuntu 14", "Ubuntu 16", and "Windows Server 2012 R2"
"Functionality"	Web browsing analysis and continuous integration tracking	Machine learning (ML), flow analysis, information extraction, and streaming	Advanced analytics, vendor support for several Big data concerns, and the creation and implementation of data mining apps	Information technology in streams and batches
Unification of services	Yes	Yes	Yes	Yes
Location of "Deployment"	"Zonal"	"Zonal"	"Regional"	"Regional"
Component of Operation	"EC2 cluster"	"Cluster"	"Cluster"	"Cluster"
Scalability of Computing operations	"Manual"	"Manual"	"Manual"	"Manual"
Pricing Structure for Compute Nodes	Hourly basis payment	Payments per second	Hourly basis payment	Minute based payments

(Continued)

TABLE 7.2 (Continued)
Frameworks for "Cloud-Based Big Data" Enterprise Solutions [25]

Sl. No.	Architecture and Operations	"AWS"	"GCP"	"IBM Cloud"	"MS Azure"
[-5ex]2	"Big Data Storage"	"Amazon S3"	"Google Cloud Storage"	"IBM Cloud Object Storage"	"Azure Blob Storage"
	Form of Memory	Archiving of Distributed Entities	Archiving of Distributed Entities	Archiving of Distributed Entities	Archiving of Distributed Entities
	Kinds of entities	"Object"	"Object"	"Object"	Blobs can be appended or blocked
	Component for Deployment	"Bucket"	"Bucket"	"Bucket"	"Container"
	Identity for Deployment	Code that is globally unique	Code that is globally unique	Unique identifier for everyone	Distinctive Identifier at the Account Level
	Scalability of Storage	Automatic scaling	Automatic scaling	Manual scaling	Manual scaling
	Location of "Deployment"	"Regional"	"Regional and multiregional"	"Regional and multiregional"	"Zonal" and "Regional"
	Metadata for Entities	"Yes"	"Yes"	"Yes"	"Yes"
	Layout for Information Storage	"Any"	"CSV, JSON (Newline Delimited Only), Google Cloud Datastore Backups, Avro"	"Any"	"XML"
	Information Storage Capacity	Unrestricted	Restricted	Unrestricted	Restricted
	Configuration Management of Entities	Yes	Yes	Yes	No
	Memory Types	"Amazon Glacier", "Standard", Irregular Connectivity	"Nearline", "Coldline", "Regional", "Multiregional"	"Standard", Crypt, as well as Cold Crypt Connectivity, Flex Category (Varying information)	Duplicate Categories: "LRS", "GRS", "ZRS", "RA-GRS" Tiers: Hot, Cool

(Continued)

TABLE 7.2 (*Continued*)
Frameworks for "Cloud-Based Big Data" Enterprise Solutions [25]

Sl. No.	Architecture and Operations	"AWS"	"GCP"	"IBM Cloud"	"MS Azure"
	Object Versioning	Yes, Automatic	Yes, Automatic	Yes, Manual (Private Cloud Implementation), No (Dedicated/public cloud implementation)	Yes, Manual
[-5ex]3	"Big Data Warehouse"	"Amazon Redshift"	"Google BigQuery"	"IBM Db2 Warehouse on Cloud"	"Azure SQL Data Warehouse"
	Component of Deployment	"Cluster" (EC2)	Completely regulated	"Cluster"	"Massively Parallel Processor" (MPP)
	Quantify Scalability	Yes, Automated	Yes, Automated	Yes, Automated	Yes, Physical
	Allowable range of Compute Nodes	1–128	NA, Fully Managed	1–60 (1 head node, 50+ information entities)	1–60
	Location of Deployment	"Region"	"Region"	"Region"	"Region"
	Format of storage	"Columnar"	"Columnar"	"Columnar"	"Columnar"
	Origins of information	"Amazon S3", "Amazon Dynamo DB", "Amazon EMR", "AWS" information Channel	"Google Cloud Storage", "Google Cloud Dataflow", readable data sources	Information file on network, information stores like Amazon S3 or IBM Cloud Object Storage, and Db2® server	Azure Blob Storage
	Techniques of Information loading	"COPY" from S3, "Streams" from "Amazon Kinesis Firhose"	"Google Analytics" Professional, Live Uploading, Batch Streaming	Read from the cloud, through a disk Gps, Twitter information, and publicly available information can all be loaded	Information loading using "PolyBase", "SQLBulkCopy" Interface, SSIS (SQL Server Integration Service) with BCP Function, with Azure Information Warehouse
	"Query Language"	"PostgreSQL"	"Standard SQL (Beta), Legacy BigQuery SQL"	"SQL Reference, CLP+, SQL PL, PL/SQL"	"PolyBase T-SQL"

7.3.3 "AZURE SQL DATA WAREHOUSE (BIG DATA WAREHOUSE SOLUTION)"

"SQL Data Warehouse" is a platform for substantially parallelization using SQL insights that allows for flexible and autonomous computation and storage scaling, as well as easy interaction utilizing big data repositories to create a center for cube and data stores. Unlimited scalability, adaptability as well as extension, high cybersecurity, and interoperability with Microsoft as well as other major manufacturers using other core technologies are among its advantages and benefits of use [30,38,39].

7.4 ASSESSMENT OF CLOUD-CENTERED "BIG DATA" BUSINESS SOLUTIONS DESIGNS

Table 7.2 lists the relative study of the summarized commodities, in which comparisons are made across relevant products within a sector, such as AWS's chosen "big data analytics" tool versus big data analytics tools from the other three companies. The objective of this comparison is not to disparage any of the cloud-based "big data" tools discussed. The goal is to show that each has a substantial application in many areas.

7.5 REAL-WORLD CASE STUDIES IN BIG DATA AND CLOUD-BASED ANALYTICS

This section discusses some relevant real-world case studies in the domain of big data and cloud-based analytics.

7.5.1 BIG DATA CASE STUDY-1 – "WALMART"

With more than 2 million workers and 20,000 locations in 28 countries, "Walmart" is the nation's biggest retail and revenue-generating enterprise. It began using "big data analytics" before the phrase "big data" was introduced. "Walmart" utilizes information analysis to discover trends that could be utilized to make product suggestions to customers based on which items were combined. "WalMart" has boosted its user exchange value by implementing effective "Data Mining". It has been increasing "big data analysis" to produce the best e-commerce technologies and improve customer service. The main aim of "Walmart's" massive information space is to improve services' retail experience while in a "Walmart store". "Walmart's" big data services were created with the objective of rebuilding international networks and creating innovative products to personalize clients' retail experiences while increasing operational efficiencies. Internally, consumers have access to real-time information recorded from many resources and organized for efficient usage using "Hadoop" and "NoSQL" technologies [41].

7.5.2 BIG DATA CASE STUDY 2 – "UBER"

Whenever it comes to moving passengers, including delivering packages, "Uber" is the best pick for millions of individuals. It utilizes the person's private data to track which parts of the business are most frequently used, analyze usage statistics, and identify areas in which the products would also be targeted more. "Uber" focuses on the market forces of goods that cause the cost of services to fluctuate. As a result, dynamic pricing is among Uber's most essential information applications. For example, if you are late for an appointment and reserve a vehicle in a congested area, you should consider paying double as much. For instance, the traveling expense of 1 mile to a festival could have a high range. Dynamic pricing and economic circumstances in the near future could be the difference between preserving and retaining consumers. Machine learning (ML) methods are used to assess where there is a large market [41].

7.5.3   BIG DATA CASE STUDY 3 – "NETFLIX"

This is the most popular American media firm, focusing on offering consumers Internet video on-demand streaming. "Netflix" has set its sights on using big data to forecast what its users will prefer to watch. As a consequence, "big data analytics" operates as the fuel for the "recommendation engine" designed for this purpose. "Netflix" has recently begun portraying itself as a content producer rather than merely a consumer. This technique has, presumably, been heavily influenced by statistics. Data elements such as what titles users view, the number of times viewing is halted, and ratings are provided are input into "Netflix's" recommendation algorithms, and fresh content choices. "Hadoop", "Hive", and "Pig" are among the firm's data structures, along with a variety of certain other standard analytics tools. "Netflix" demonstrates that understanding precisely what customers want is simple if businesses do not even assume and instead use big data to make decisions.

7.5.4   BIG DATA CASE STUDY 4 – "EBAY"

As a data-intensive company, "eBay" faces a significant technological barrier in implementing measures that could quickly analyze and process the information as it emerges (streaming data). Several ways of supporting stream processing interpretation have been developed rapidly. "eBay" uses a variety of tools, notably "Apache Spark", "Storm", and "Kafka". It enables the firm's technical experts to explore for informational markers connected with the information (schema) and allow access to as many individuals as possible while maintaining the additional level of protection and privileges. The firm is now at the center of things for implementing big data technologies and plays a significant role in the open-source community [41].

7.5.5   BIG DATA CASE STUDY 5 – "PROCTER & GAMBLE"

"Procter & Gamble" (P&G) is a 170+-year-old corporation providing commodities used by many. The brilliant corporation saw the potential of big data and applied it in functional departments globally. P&G has placed a heavy focus on leveraging big data to make better, faster company choices. International Services Companies have created products, techniques, and procedures that cross organizational boundaries and provide instant access to the most up-to-date data with advanced insights.

As a result, despite numerous new firms, P&G, as the oldest, continues to have a large market share [41].

7.5.6 CASE STUDY 6: BIG DATA PREDICTING THE UNCERTAINTIES

A pioneering study in Bangladesh discovered that tracking movements of people throughout the region utilizing information via mobile telephony can assist anticipate where infectious diseases like malaria are likely to happen, allowing healthcare companies to take precautionary actions. "Malaria" kills around 4 lakh people worldwide annually, the majority of which are youngsters. The information, such as details from the Bangladesh health ministry, is being used to generate risk maps, which exhibit the likely areas of "malaria" outbreaks, letting regional health authorities get workable safeguards like the spritzing pesticides and herbicides and amassing mosquito nets as well as medications to safeguard the population from the illness [41].

7.5.7 DEPLOYMENT OF "BIG DATA" FOR AD NETWORK ASSESSMENT IN MORE THAN TEN REGIONS OF THE WORLD [42]

In this case study, the client was a well-known market research firm. The main issue was that, although possessing a reliable statistical platform, the client feared this would not be able to meet the firm's expected demand. Recognizing the predicament, the client had been on the lookout for a forward-thinking, inventive remedy. The aim of this system was to deal with the ever-increasing volumes of information, analyze large information quickly, and provide full marketing network analytics. The client was looking for a professional and experienced group to execute the change after settling on the platform's design. After a lengthy relationship with "ScienceSoft", the client asked the specialists to handle the whole transfer from the legacy to the new analytical model. The solution for the above is summarized as follows. The client's predictive analytics designers teamed up with "ScienceSoft's big data" group throughout the work. The client team created an idea, while the "ScienceSoft" team was in charge of putting it into action. The client's designers chose the ensuing components for the new analytical mechanism:

- pache Hadoop" – for information storage.
- pache Hive" – for consolidation, querying, and assessment of information.
- pache Spark" – for information processing.
- As "cloud computing systems", "Amazon Web Services", and "Microsoft Azure" were chosen.

At the client's request, the old system and the newer one ran in tandem during the transition. The following important components were incorporated into the actual approach:

- Compilation of information.
- taging".
- ata warehouse-1".

- ata warehouse-2".
- esktop application".

Collected information was sent to the platform from a multitude of different sources, including television viewing, portable phone browser habits, and Internet traffic statistics, as well as questionnaires. Information gathering included the following Python-coded phases to allow the technology to handle over 1,000 distinct categories of original information (cache, Excel, Text, and so forth):

- Information transition.
- Processing of information.
- Information fusion.
- Information uploaded into the database.
- taging": the foundation of such a component was "Apache Hive". The information format was comparable to raw database schema at the time, and there were no documented links among participants from various channels, such as television and the Internet.
- ata warehouse-1": this block is built on "Apache Hive", the same as the preceding one. Information mapping occurred here. As per the modeling guidelines, the computer analyzed the responders' information for audio, television, web, and paper resources, and connected consumers' IDs across multiple data sources. "Python" was used to write the ETL for that component.
- ata warehouse-2": utilizing "Apache Hive" and "Spark" at its foundation, the module ensured real-time data analytics based on business rationale: summation, averaging, likelihood, and so on. The desktop app's SQL statements were processed using "Spark's" Data Frames. "Scala" was used to create ETL. "Spark" also enabled exhaustive data to be filtered based on the program's customers' authorized access.
- esktop application": the new approach allowed for cross-analysis of nearly half a million features and the creation of intersecting vectors that allowed for cross business intelligence from multiple sectors. The client could produce informal results in combination with conventional statistics. The platform provided a speedy reply of convenient graphs once the client picked numerous factors of relevance (for instance, a specific television network, a class of consumers, or a specific time of day). The prediction may also assist the consumer. The program, for instance, might anticipate sales compared to projected reach and anticipated advertising expenditure.
- Findings: the proposed program could handle some requests approximately 100 times quicker than the previous approach at the program's conclusion. The client was enabled to do complete promotional campaign research for several marketplaces using the insightful information gained from the study of around 1 million parameters.
- Systems and Applications: "Apache Hadoop", "Apache Hive", "Apache Spark", "Python (ETL)", "Scala (Spark, ETL)", "SQL (ETL)", "Amazon Web Services (cloud storage)", "Microsoft Azure (cloud storage)", ".NET (desktop application)".

7.5.8 CASE STUDY 8: DESIGN OF AN IoT PET MONITOR USING BIG DATA STRATEGY

In this case study, the client was a corporation located in the United States that operates in 18 states. Having extensive telecom expertise, the client chose to develop a new platform that allows pet parents to follow their animals' movements using wearables sensors controlled through a smartphone app. The client's objective was to find a big data solution, which might help customers to constantly know where their animals were, receive real-time warnings regarding key occurrences, and obtain statistics on their animals' whereabouts. The idea was to allow for the sharing of audio-visual content (voice, film, and images) such that animal parents may communicate with their animals and know where they have been at any given time.

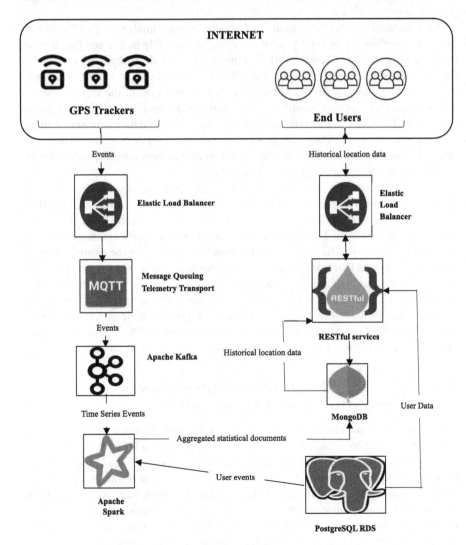

FIGURE 7.3 Big data deployment on the web.

Because the client anticipated the number of customers to continue to rise, the system needed to be highly adaptable to collect and organize more content [43]. To make sure that the system is easily customizable, "ScienceSoft's" big data team deployed it on the web and built it around "Apache Kafka", "Apache Spark", and "MongoDB" (Figure 7.3).

Utilizing the "MQTT" protocol, many global positioning send actual information on pet whereabouts as well as occurrences (for example, low power and exiting a secure zone) toward the mail server. The standard was selected since it ensures a gadget-friendly interface while also saving battery capacity on cell devices. An "Apache Kafka"-based dynamic data synthesizer pulls information from several "MQTT" components, analyzes it instantaneously, and verifies the quality of content. The "Kafka Streams" module enables push alerts and assures secure information transport.

An information integrator built on "Apache Spark" analyzes stored information, then combines it by hourly, daily, weekly, and monthly before sending it to an information store. The "ScienceSoft's" team recommended "MongoDB" knowledge for the latter since it enables time-series data occurrences to be stored as a single document on a daily, hourly, and monthly basis. Furthermore, the report-oriented design permitted in-place modifications, resulting in significant efficiency gains. Customers' identities, credentials, and administrative information were stored in the existing repository on "PostgreSQL RDS". "RESTful" services decouple the user interface from the information storage and also ensure dependability, adaptability, and system or coding environment independence.

The client got a robust big data solution, which could manage 30,000+ transactions every moment across 1 million sensors. Therefore, users could monitor their pets' whereabouts instantaneously as well as transmit and receive images, movies, and voicemail messages. The consumer gets text alerts if a significant incident occurs (for example, a pet breached a geographical boundary created by the animal parent or the animal's wearable tracker went "out of communication"). Animal holders could likewise view hourly, quarterly, or periodic statements that are generated routinely, or adjust the recording phase individually if necessary. The technology devices used in the project were: "Amazon Web Services", "MQTT", "Apache Kafka" (stream data processor), "Apache Spark" (data aggregator), "MongoDB" (data warehouse), "PostgreSQL RDS" (operational database), and "RESTful" web services.

7.5.9   CASE STUDY 9: DEVELOPMENT OF A TELECOMMUNICATION COMPANY'S DATA ANALYTICS SOLUTION [44]

In this case study, the client was a Texas-based telecommunication organization that participated in the government Lifeline Recovery Program and provided low-income persons with pre-paid cellular phones with service packages. The "ScienceSoft's" analytics group was tasked with designing and implementing a data processing and analytics tool that would allow the client to gather data from numerous sources and gain insights into consumer behavior. The system had to be able to examine historic information and analysis. An additional challenge to resolve was access permissions, as the client intended to give their renters exposure to rental statistics.

Actual data (including user views and buttons, pricing packages, screen sizes, and applications loaded) were collected across 10+ providers by the data analytics solution. The "MQTT" protocol was proposed by "ScienceSoft's" team to capture the sensor information as well as transfer it into "Apache Kafka". To save money on AWS computational power, the team recommended adopting "Amazon" Spot Instances. Researchers deployed AWS Application Proxy Balancers to guarantee the analytic game's adaptability.

"Apache Kafka" served as a medium for streaming data. The gathered information was sorted there before being offloaded onto the "Amazon Simple Storage" Service landing zone. "Amazon Redshift" was selected for information warehousing and distribution, wherein telemetry information from Android smartphones, as well as statistics from "Enterprise Resource Planning" and the "Home Location Register" (HLR), were sent.

The "Science Soft" created ROLAP modules with 30+ measurements and 10+ statistics to facilitate periodic and ad-hoc analysis. For example, the analytic system provided an overview of the loyalty points gained by measuring a user's advertisement exposures as well as click-throughs. The client might infer that the client was unsatisfied with the product centered on the enhanced total of requests to assist. Without action, this might result in client attrition.

The system was made available not only to the clients but also to their renters (including telecommunication firms with their own clients and HLRs). A renter, for example, could view the portion of statistics that pertains to their business. The "ScienceSoft's" team devised two methods to do this: public entry (managed at the data storage tier) and exclusive accessibility using a dedicated AWS account. The client was able to demonstrate the ability with "ScienceSoft's" big data solutions as follows:

- Determine how engaged a customer is and what their interests are.
- Recognize patterns in consumer's behavior.
- Do forecasting regarding how consumers will act.
- Marketers will be invoiced depending on their determined proportion.
- Employ predictive analytics to your advantage. For instance, current earnings, the rate of new subscribers, and customer support statistics, among other things.
- The client was able to save 80% on AWS computational resources by using "Amazon Spot Instances".

The tools used in this case study were "Amazon Web Services (Amazon cloud)", "Apache Kafka" for streaming live information, the "Message Queuing Telemetry Transport" Protocol, "Amazon Simple Storage Service" for continuous storage, "Amazon Redshift" as a "data warehouse", "Airbnb Airflow", and "Python (ETL)".

7.5.10 CASE STUDY 10: IMPLEMENTATION OF DATA ANALYTICS FOR A MULTIBUSINESS ORGANIZATION [45]

The client in this case study was an American-based firm operating a multichannel commerce, lodging, cafe, and other enterprises. The client's key goal was to increase client retention through a personalized technique while also optimizing intrinsic retail operations. They could not do it, though, because the information was locked away in numerous programs that were specialized to their operations strategy.

Because the software's goal was to satisfy all of the company's business instructions by collecting and aggregating information from 10+ information resources, including "CRM", "Magento", and "Google Analytics", as well as devoted restaurant, eatery, and wellness mechanisms, "ScienceSoft's" team initially supplied the client with a prototype and then formulated a series of test reports centered on the client's ERP statistics.

The group at "ScienceSoft" developed high-level design modules along with their primary tasks. The analytic solution is available with scalability in mind. The goal was to evaluate historical information for 5 years at first, and then to handle information expansion in the future. Because the client was concerned about the security of information stored in a private cloud in a data center, the approach was blended.

"ScienceSoft" suggested a technological architecture, which would meet the client's solution requirements, like adaptability, speed, and dependability both for desktop and mobile consumers. They evaluated whether these technologies and the accompanying "Microsoft stack" were viable for the possible answer because the client always had certain old systems operating on "Microsoft SQL Server". This might enable the consumers to cut deployment expenses as only a few extra licenses would be needed.

"ScienceSoft" additionally recommended that the workaround be enhanced with sophisticated analytics abilities depending on user experience of the retail industry. They produced a prototype for a recommender system as the forecasting algorithm underlying the machine was to increase the user's cross-selling and the up-selling prospects for the virtual retailer, and a "time-series" projection prototype to estimate profits, for that.

The elements of the developed analytical solution were as follows:

- A data center that stores regular and semi-structured information from 15 diverse information sources.
- ETL (extract, transform, and load) operations number in the hundreds.
- A "data warehouse" to merge and integrate information.
- A statistical system with five OLAP cubes and around 60 parameters.
- Generation of reports.
- To guarantee that the client may gain through intermediate outputs, the integration was broken into many phases. They created around 90 modules for the client's various enterprise directions and consumer management in total. Because the information assimilation from different systems is worthless without a well-defined information continual improvement procedure, The "ScienceSoft's" team devised ETL guidelines that were designed to
  - Combine master information from several systems, such as client profiles.
  - Consolidate information into a single format.

The client's information management procedure was substantially automated thanks to these guidelines. Administrative manipulations by an information administrator, on the other hand, were mostly conceivable.

"ScienceSoft" also expanded on access control restrictions to maintain data protection. Just before the program started, they examined the client's extremely customizable access model and determined that this would be unacceptable, since it would adversely impair the statistical optimizer efficiency (i.e., the system could take too much time to generate the needed statistics). As a result, they suggested a less sophisticated but still very effective three-level access approach (for a particular enterprise, division, or personnel). The ideas and innovations had no detrimental effects on the program's operation.

"ScienceSoft's" staff offered extensive assistance to the client as part of the supplied business intelligence solutions. For instance, once the client's third-party analytical suppliers made some adjustments, they offered training on installing and dealing with "OLAP cubes" and updated "ETL" procedures.

The client benefitted from a 360° consumer point of view throughout all channels and company objectives, and strong retail intelligence, which helped in creating a personalized customer experience, thanks to the built intelligence platform. By enhancing inventory levels and evaluating employees, the client also was able to optimize the inner company's operations.

After integrating all the information, the client was empowered to

- Examine the consumers' purchasing behavior and patterns.
- Examine the customers' recent behavior, regularity, and financial worth.
- Determine their most important consumers.

The following data were analyzed by the client:

- The most/least frequented sites, sites with really no traffic, and pages with traffic jams and low conversions are all examples of traffic and conversion percentages.
- Participation of Internet shopping consumers.
- Merchandise wish lists, revenues, as well as basket dumping.

Rather than relying on a data repository to show the inventory levels and the need for continual corrections over the telephone, the client was able to monitor the real stock position at both the facility and the stores in real time. Such inventory stock visibility has had a favorable impact on the purchasing and logistical operations. The client was able to define the overall quality of the workers' activity using KPIs and goal management reports. "Microsoft SQL Server", "Microsoft SQL Analysis" and "Integration Services", "Python", and "Microsoft Power BI" were used.

7.5.11   CASE STUDY 11: BIG DATA CONSULTING AND TRAINING FOR A SATELLITE AGENCY [46]

The client in this case study was a geospatial organization that gathers and analyzes meteorological and environmental data. The difficulty was that the client has a well-planned training program, pursuant whereby the organization intends to start a new project involving satellites geosynchronous. They would be gathering and relaying meteorological, temperature, and ecological survey data. To meet the deadline, the client hired a contractor ahead of time to develop the structure of the future solution and select the software system. Because the program is designed to last three decades and having the right design is critical, the client opted to hire other firms to review the documents that have been acquired and selected "ScienceSoft" for the job. Those who also asked "ScienceSoft" to develop and execute training courses on the big data environment in general and the recommended solutions in specific, as the workaround depends on big data.

"ScienceSoft's" big data consultancy group starts by visiting the client on website to know about their company and the product they were considering. They also held group discussions with client's staff. Our specialists then dug into the offered information on the recommended design and computing environment. In much less than a week, our crew had to consider carefully 1,000+ pages, highlight the good aspects, and, if necessary, bring out whatever omissions or flaws.

The staff at "ScienceSoft" has created educational materials that will be provided over the next initial appointment. The experts provide a summary of current big data technologies to expose the client to the big data ecosystem going far beyond those recommended by the third-party contractor. The presentation also includes application scenarios, with an emphasis on those that would be relevant to the client's industry. The experts designed entirely different sets of learning resources for every potential customer because "ScienceSoft's" team was really to conduct training courses both for senior knowledge and professional staff.

The "ScienceSoft's" team was prepared not only to lead seminars but also to create experience and understanding resources and offer advice on how to evaluate test takers. Nevertheless, the client preferred a conference approach with Q&A sessions. Besides that, "ScienceSoft" offered technical workers with on-the-job assistance, with the experts conducting coding workshops during the second onsite visit.

The client gets an assessment report detailing the optimizer's strengths and weaknesses. The study also included suggestions for how to enhance the system. For instance, substituting one of the recommended technologies with the other resulted in improved quality and durability for the system. "ScienceSoft" has created over 300+ slides of training manuals that present the big data idea, technology, and real-world experiences. NoSQL technology, and broadcasting and queuing analysis, were one of the subjects addressed. Seminars with panel discussions and on-the-job help were used to practice regularly. After the session, the client gets all of the documents, which became their intangible asset.

Q&A meetings, quantitative performance evaluations (accessibility, integration, expandability, serviceability, ease of maintenance, vendor support, and achievement), and remedy evaluations related to technological features (consumer experience, technology maturity, digital innovation, and supplier assistance) were some of the methodologies used in this case study.

7.5.12   CASE STUDY 12: BIG DATA CONSULTING FOR A LEADING INTERNET OF VEHICLES COMPANY [47]

The client is a transportation and investment management company, which is one of the biggest Internet-of-things firms in the European Union (EU). To allow customers of their services to make informed decisions, the client gathers, saves, and analyzes IoT data from 1 million automobiles linked to their platforms. The client sought to enhance their big data analytics abilities after implementing a comprehensive IoT information gathering and storage system. "Apache Cassandra" is used in their main approach. "Apache Cassandra", described as the key, does not build sustainable large amounts of data analytics very well. To download information for 70,000 vehicles, for instance, the client must dynamic responses 70,000 distinct queries, each of which will receive 70,000 unique reports, necessitating more attempts to obtain a comprehensive picture.

The big data specialists visited the client for a 3-day onsite visit, where they thoroughly analyzed the existing schema, including its design, relevant documentation, current evidence streams, and existing data standard operating procedures. Following that, we organized a workshop devoted to the future solution, wherein they addressed the optimizer's estimated launch date, current and chosen technology, and accessible licenses. They additionally established structurally required specifications for the optimizer's reliability, speed, privacy, and adaptability during the session. Next, they classified every demand as crucial, moderate, mid, or minor in terms of its influence on the firm's success.

The experts created the design concept for the future solution based on the conclusions of the session. They discussed the responsibilities of high-level design modules such as continuous integration as well as a cloud service. Following the presentation of the design approach, the experts conducted a Conversation during which they offered definitive solutions to any questions the client had. For example, they discussed the benefits and drawbacks of a datastore vs a centralized data warehouse. They further contrasted on-premises deployment with cloud-based solutions, covering several public scenarios – including "Amazon Web Services" and "Microsoft Azure" at the forefront. Furthermore, the "Apache Cassandra" consultants gave the client ideas about how to increase "Apache Cassandra's" performance, including table layout, partitioning variables, and the type of information that can be stored.

The following information was included in the client's visit document:

- Architecturally significant criteria for the future solution, as well as their business impact. The responsibilities of essential architectures are outlined at a high level in this high-level design.
- The client gets detailed responses to their inquiries about the prospective solution's deployment, the benefits and drawbacks of various technologies and architectural aspects, the complexities of disk space and connection, and other key topics. Analysis of business needs, function breakdown, workshop on structurally critical aspects, Q&A session, and technology comparison.

7.6 CHALLENGES AND ISSUES

Certain hurdles must be addressed in order to progress above current ML and big data approaches and methods. The following criteria are viewed as crucial by NESSI [20].

- A real scientific foundation must be established before selecting an appropriate approach or concept.

- Novel scalable and efficient methods must be created.
- Suitable development abilities and technology frameworks must be developed and implemented to properly apply conceived remedies.
- Finally, the solutions' value creation must be investigated just as thoroughly as the data format and accessibility.

Ongoing problems, such as acceptance and application of quality big data solutions employing cloud infrastructure, as well as addressing vulnerabilities, arise in the context of virtualized "big data analytics". One of the most pressing challenges when combining big data analytics with cloud technology is security. This is why cloud-based big data analytics, as well as its application in practice and execution, has gotten so much interest.

The researchers summarized, assessed, and compared authenticator-based information integrity testing strategies for "cloud" and "Internet-of-things" data [48]. Any further advancements in this field, according to this article, should consider three major facets: effectiveness, reliability, and expandability. On "cloud" and "Internet-of-things" information, the authors [48] give an overview, evaluation, and comparison of authenticator-based file integrity verification methodologies. Any future advancements in this field, according to this article, should consider three primary aspects: efficiency, reliability, and expandability.

The authors [49] advise that the following topics must be investigated further:

- Hierarchical programming or extensible high-level methodology.
- Interchange of information and technological solutions.
- Combining several big data analytics platforms.
- Proprietary data mining methods.

In a decentralized computing environment known as fog computing, data, compute, storage, and applications are spread between the information source and the cloud. A decentralized computing system called edge computing puts enterprise applications closer to sources of data like IoT gadgets or regional edge servers. Strong commercial advantages, including quicker insights, quicker response times, and greater bandwidth availability, can result from being close to the source of the information. Durability, complexity, dynamicity, diversity, delay, and privacy are a few issues with fog computing that are related to big data. The issues of edge computing related to big data include storage costs, lost information, and privacy.

Big data peregrination is a long journey. It involves network hurdles and interdisciplinary interactions, dealing with expensive data to name a few [53].

7.7 CONCLUSION

The advent of this subject of study has grabbed the attention of several professionals and scholars in this age of big data. Big data analytics and analysis have grown increasingly important as the pace at which information is generated in the virtual environment has increased. Furthermore, most of this information is already on the cloud. As a result, relocating "big data analytics" to the "cloud" is a feasible choice. Furthermore, the cloud structure meets the data analytics techniques' storing and processing needs. On the other hand, there are conflicts such as safety, confidentiality, and a loss of ownership and control. This chapter surveyed some of the important aspects of intelligent analytics in big data and cloud. Several significant tools in "big data" and "cloud" are compared based on their performance. Numerous real-world case studies (12 in particular) on intelligent analytics in big data and cloud are discussed in detail. The challenges and issues faced while developing big data analytics in cloud along with future research directions are summarized in this chapter.

7.8 FUTURE RESEARCH DIRECTIONS

There are a variety of open-source data mining approaches, tools, and applications. "R", "Gate", "Rapid-Miner", and "Weka" are just a few of the tools available. "Cloud-based big data analytics" solutions must have the permission to connect these cheaper data analyses on the cloud system to provide expense and quality service. The availability, expense, and convenience of setup and evaluation of cloud-based statistics are the primary reasons for its popularity. Considering this, researchers [50] outlined the following important study guidelines:

• Statistics and data management have evolved in relation to cloud-based analytics.
• Adapting and evolving approaches and approaches to boost efficiencies and minimize hazards.
• Develop plans and tactics to address security and privacy issues.
• Examining and adapting ethical and legal practices to reflect shifting perspectives, impacts, and consequences of technological advancements in this domain.

The study possibilities, however, really are not confined to the aforementioned points. The primary objective would be to shift the cloud from an information management and infrastructure base to a flexible and scalable analytics solution.

REFERENCES

1. Khan, S., Shakil, K. A., and Alam, M. (2008). Cloud-Based Big Data Analytics-A Survey of Current Research and Future Directions. *Advances in Intelligent Systems and Computing, 654*. https://doi.org/10.1007/978-981-10-6620-7-57
2. Chen, C. L. P. and Zhang, C. Y. (2014). Data-Intensive Applications, Challenges, Techniques, and Technologies: A Survey on Big Data. Information Sciences. *Advances in Intelligent Systems and Computing, 275*, 314–347. https://doi.org/10.1016/j.ins.2014.01.015
3. Driscoll, A., Daugelaite, J., and Sleator, R. D. (2013). 'Big Data', Hadoop and Cloud Computing in Genomics. *Journal of Biomedical Informatics, 46*(5), 774–781. https://doi.org/10.1016/j.jbi.2013.07.001
4. Manekar, A. and Pradeepini, G. (2015). A Review on Cloud-based Big Data Analytics. *ICSES Journal on Computer Networks and Communication, 1*(1), 1–4.

5. Assuncao, M. D., Calheiros, R. N., Bianchi, S., and Netto, M. A. S. (2015). Big Data Computing and Clouds: Trends and Future Directions. *Journal of Parallel and Distributed Computing, 79,* 3–15. https://doi.org/10.1016/j.jpdc.2014.08.003

6. Wu, X., Zhu, X., Wu, G.-Q., and Ding, W. (2014). Data Mining with Big Data. *IEEE Transactions on Knowledge and Data Engineering, 26*(1), 97–107. https://doi.org/10.1109/TKDE.2013.109

7. Pashazadeh, A. and Jafari Navimipour, N. (2018). Big Data Handling Mechanisms in the Healthcare Applications: A Comprehensive and Systematic Literature Review. *Journal of Biomedical Informatics, 82,* 47–62. https://doi.org/10.1016/j.jbi.2018.03.014

8. Challita, S., Zalila, F., Gourdin, C., and Merle, P. (2018). A Precise Model for Google Cloud Platform. *IEEE International Conference on Cloud Engineering,* 177–183. https://doi.org/10.1109/IC2E.2018.00041

9. Shakil, K. A., Sethi, S., and Alam, M. (2015). An Effective Framework for Managing University Data Using a Cloud Based Environment. *2nd International Conference on Computing for Sustainable Global Development (INDIACom),* 1262–1266.

10. Jackson, J. C., Vijayakumar, V., Quadir, M. A., and Bharathi, C. (2015). Survey on Programming Models and Environments for Cluster, Cloud and Grid Computing That Defends Big Data. *2nd International Symposium on Big Data and Cloud Computing,* 517–523.

11. Neaga, I. and Hao, Y. (2014). A Holistic Analysis of Cloud Based Big Data Mining. *International Journal of Knowledge, Innovation and Entrepreneurship, 2*(2), 56–64.

12. Borthakur, D., Gray, J., Sarma, J. S., Muthukkaruppan, K., Spiegelberg, N., Kuang, H., Ranganathan, K., Molkov, D., Menon, A., Rash, S., Schmidt, R., and Aiyer, A. (2011). Apache Hadoop Goes Real-Time at Facebook. *ACM SIGMOD International Conference on Management of Data,* 1071–1080.

13. Strambei, C. (2012). OLAP Services on Cloud Architecture. *Journal of Software and Systems Development.* https://doi.org/10.5171/2012.840273

14. Khan, I., Naqvi, S. K., Alam, M., and Rizvi, S. N. A. (2015). Data Model for Big Data in Cloud Environment. *Computing for Sustainable Global Development (INDIACom),* 582–585.

15. Ortiz, J. L. R., Oneto, L., and Anguita, D. (2015). Big Data Analytics in the Cloud: Spark on Hadoop vs MPI/OpenMP. *INNS Conference on Big Data, 53,* 121–130.

16. Baker, T., Al-Dawsari, B., Tawfik, H., Reid, D., and Nyogo, Y. (2015). GreeDi: An Energy Efficient Routing Algorithm for Big Data on Cloud. *Ad Hoc Networks, 35,* 1–14.

17. Li, X., Calheiros, R. N., Lu, S., Wang, L., Palit, H., Zheng, Q., and Buyya, R. (2012). Design and Development of an Adaptive Workflow-Enabled Spatial-Temporal Analytics Framework. *18th International Conference on Parallel and Distributed Systems,* 862–867.

18. Calheiros, R. N., Vecchiola, C., Karunamoorthy, D., and Buyya, R. (2012). The Aneka Platform and QoS-Driven Resource Provisioning for Elastic Applications on Hybrid Clouds. *Future Generation Computer Systems, 28*(6), 861–870.

19. Kim, H., Abdelbaky, M., and Parashar, M. (2009). CometPortal: A Portal for Online Risk Analytics Using CometCloud. *17th International Conference on Computer Theory and Applications.*

20. Chen, Q., Hsu, M., and Zeller, H. (2011). Experience in Continuous analytics as a Service (CaaaS). *14th International Conference on Extending Database Technology,* 509–514.

21. Srivastava, M. and Yadav, P. (2021). Build a Log Analytic Solution on AWS. *5th International Conference on Information Systems and Computer Networks,* 1–5. https://doi.org/10.1109/ISCON52037.2021.9702374

22. Neumeyer, L., Robbins, B., Nair, A., and Kesari, A. (2010). S4: Distributed Stream Computing Platform. *IEEE International Conference on Data Mining, ICDM,* 170–177. https://doi.org/10.1109/ICDMW.2010.172

23. Biem, A., Bouillet, E., Feng, H., Ranganathan, A., Riabov, A., Verscheure, O., Koutsopoulos, H., and Moran, C. (2010). IBM InfoSphere Streams for Scalable, Real-Time, Intelligent Transportation Services. *ACM SIGMOD International Conference on Management of Data,* 1093–1104. https://doi.org/10.1145/1807167.1807291

24. Zhang, Z., Liu, Z., Jiang, Q., Chen, J., and An, H. (2021). RDMA-Based Apache Storm for High-Performance Stream Data Processing. *International Journal of Parallel Programming, 49,* 671–684. https://doi.org/10.1007/s10766-021-00696-0.

25. Saif, S. and Wazir, S. (2018). Performance Analysis of Big Data and Cloud Computing Techniques: A Survey. *Procedia Computer Science, 132,* 118–127. https://doi.org/10.1016/j.procs.2018.05.172

26. Kumar, V. D. A., Divakar, H., and Gokul, R. (2017). Cloud Enabled Media Streaming Using Amazon Web Services. *IEEE International Conference on Smart Technologies and Management for Computing, Communication, Controls, Energy and Materials (ICSTM), 132,* 118–127. https://doi.org/10.1016/j.procs.2018.05.172

27. Gulabani, S. (2017). Practical Amazon EC2, SQS, Kinesis, and S3 A Hands-On Approach to AWS. *eBook: Springer.* https://doi.org/10.1007/978-1-4842-2841-8

28. Pradhananga, Y., Karande, S., and Karande, C. (2016). High Performance Analytics of Bigdata with Dynamic and Optimized Hadoop Cluster. *International Conference on Advanced Communication Control and Computing Technologies (ICACCCT),* 715–720. https://doi.org/10.1109/ICACCCT.2016.7831733

29. Dawelbeit, O. and McCrindle, R. (2014). A Novel Cloud Based Elastic Framework for Big Data Preprocessing. *6th Computer Science and Electronic Engineering Conference (CEEC),* 23–28. https://doi.org/10.1109/CEEC.2014.6958549

30. Ugia Gonzalez, J., and Krishnan, S. P. T (2015). Building Your Next Big Thing with Google Cloud Platform: A Guide for Developers and Enterprise Architects. *eBook: Springer.* https://doi.org/10.1007/978-1-4842-1004-8

31. Ramesh, B. (2015). Big Data Architecture. In: Mohanty, H., Bhuyan, P., and Chenthati, D. (eds) *Big Data. Studies in Big Data, 11.* https://doi.org/10.1007/978-81-322-2494-5_2

32. Patil, A., Rangarao, D., Seipp, H., Lasota, M., Marcelo dos Santos, R., Markovic, R., Casey, S., Bollers, S., Gucer, V., Lin, A., Richardson, C., Rios, R., Van Alstine, R., and Medlin, T. (2020). Cloud Object Storage as a Service: IBM Cloud Object Storage from Theory to Practice - For developers, IT architects and IT specialists. *IBM Redbooks.* ISBN: 9780738442457

33. Das, N. S., Usmani, M., and Jain, S. (2015). Implementation and Performance Evaluation of Sentiment Analysis Web Application in Cloud Computing Using IBM Blue mix. *International Conference on Computing, Communication & Automation,* 668–673. https://doi.org/10.1109/CCAA.2015.7148458

34. Nakhimovsky, A. and Myers, T. (2004). Google, Amazon, and Beyond: Creating and Consuming Web Services. *eBook: Springer.* https://doi.org/10.1109/CCAA.2015.7148458

35. Krishnan, S. P. T. and Ugia Gonzalez, J. L. (2015). Building Your Next Big Thing with Google Cloud Platform: A Guide for Developers and Enterprise Architects *eBook: Springer.* https://doi.org/10.1007/978-1-4842-1004-8_10

36. Serrano, N., Gallardo, G., and Hernantes, J. (2015). Infrastructure as a Service and Cloud Technologies. *IEEE Software, 32,2,* 30–36. https://doi.org/10.1109/MS.2015.43

37. Shovic, J. C. (2021). Connecting an IoT Device to a Cloud Server: IoTPulse. *Raspberry Pi IoT Projects.* Apress, Berkeley, CA. https://doi.org/10.1007/978-1-4842-6911-4_5

38. Soh, J., Copeland, M., Puca, A., and Harris, M. (2020). Microsoft Azure. *Apress Berkeley,* https://doi.org/10.1007/978-1-4842-5958-0

39. Moemeka, E. (2019). *Azure in the Enterprise: Cloud Architecture, Patterns, and Microservices with Azure PaaS and IaaS.* Springer.

40. Reagan (2018). *Web Applications on Azure: Developing for Global Scale.* Springer. https://doi.org/10.1007/978-1-4842-2976-7

41. 5 Big Data Case Studies – How big companies use Big Data. Retrieved from https://data-flair.training/blogs/big-data-case-studies/ on May 2, 2022.

42. Big Data Implementation for Advertising Channel Analysis in 10+ Countries. Retrieved from https://www.scnsoft.com/case-studies/big-data-implementation-for-advertising-channel-analysis on May 3, 2022.

43. Development of a Big Data Solution for IoT Pet Trackers. Retrieved from https://www.scnsoft.com/case-studies/big-data-solution-for-iot-pet-trackers on May 3, 2022

44. Implementation of a Data Analytics Platform for a Telecom Company. Retrieved from https://www.scnsoft.com/case-studies/data-analytics-platform-for-a-us-telecom-company-operating-in-18-states on May 4, 2022.

45. Data Analytics Implementation for a Multi-business Corporation. Retrieved from https://www.scnsoft.com/case-studies/data-analytics-implementation-for-a-multibusiness-corporation on May 5, 2022.

46. Big Data Consulting and Training for a Satellite Agency. Retrieved from https://www.scnsoft.com/case-studies/big-data-consulting-and-training-for-a-satellite-agency on May 5, 2022.

47. Liu, C., Yang, C., Zhang, X., and Chen, J. (2015). External Integrity Verification for Outsourced Big Data in cloud and IoT: A Big Picture. *Future Generation Computer System, 49,* 58–67.

48. Talia, D. (2013). Clouds for Scalable Big Data Analytics. *IEEE Computer Society, 46*(5), 58–67. https://doi.org/10.1109/MC.2013.162

49. Neaga, I. and Hao, Y. (2014). A Holistic Analysis of Cloud Based Big Data Mining. *International Journal of Knowledge, Innovation and Entrepreneurship, 2*(2), 56–64.

50. Saha, M., Panda, S.K., and Panigrahi, S (2021). A Hybrid Multi-Criteria Decision Making Algorithm for Cloud Service Selection. *International Journal of Information Technology, 13,* 1417–1422. https://doi.org/10.1007/s41870-021-00716-9

51. Saha, M., Kumar Panda, S., and Panigrahi, S. (2022). A Survey on Applications of Multi-Attribute Decision Making Algorithms in Cloud Computing. *ECS Transactions, 107*(1), 12887–12900.

52. Pandit, P. (2021). Case Study on AWS Identity and User Management. https://doi.org/10.13140/RG.2.2.18227.55844.

53. Behura, A. and Kumar Panda, S. (2022). Role of Machine Learning in Big Data Peregrination. Handbook of Research for Big Data, pages 42.

8 Various Audio Classification Models for Automatic Speaker Verification System in Industry 4.0

Sanil Joshi, Mohit Dua, and Shelza Dua
National Institute of Technology

CONTENTS

8.1 INTRODUCTION

Automatic speaker verification (ASV) technologies have advanced to the point that many industries and businesses such as banks are attracted in using them in actual security systems. However, the susceptibility of these systems to a variety of direct

DOI: 10.1201/9781003321149-8

and indirect access threats reduces the effectiveness of the ASV authentication mechanism. ASV system provides the verification and the authorization to the original users of the system [1].

Like other security systems, these ASV systems are also prone to various spoofing threats. With respect to ASV system, spoofing is the security attack in which imposter gain access to the original user's private information by creating the audio sample similar to the voice of the original user. In spoofing, different types of spoofing threats like replay attacks, voice conversion (VC), speech synthesis (SS), and mimicry attacks are very popular. Speech synthesis [2] and voice conversion [3] are the types of logical access attacks and are performed using advanced algorithms and functions. These attacks are performed by skilled person with deep knowledge in speech processing.

The replay attacks [4] are the simplest form of spoofing attacks. In order to perform these attacks, we just require the recording and replay devices. With the help of recording device, the attacker records the voice of the original user, and with the help of replay device, the attacker gets access to the user's private system after performing verification. These replay attacks can be performed by any person because no technical expertise is required to perform these attacks. Another important spoofing attack is mimicry attacks [5], and these attacks can be performed by skilled voice mimicry artists. Mimicry artist mimics the voice of the original user and then performs spoofing attack using mimic audio. Figure 8.1 shows the important components of ASV system.

The ASV systems comprise three main phases. First important phase is front end feature extraction phase. In feature extraction phase, the important and useful information is extracted from the input audio signal to produce the audio feature vector. Second important phase is back-end classification model. The extracted feature vector is then fed to classification model to classify the audio samples into bonafide (original) or the spoofed (cloned) audio sample. Another important phase of ASV system is the dataset because for speaker verification tasks, various speech datasets such as ASVspoof 2015 [6], ASVspoof 2017 [7], ASVspoof 2019 [8], and voice spoofing detection corpus (VSDC) [9] datasets are available.

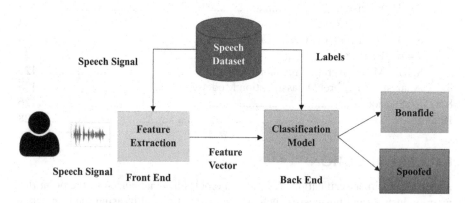

FIGURE 8.1 Components in automatic speaker verification (ASV) system.

Various feature extraction techniques such as mel frequency cepstral coefficients (MFCCs) [10], Gammatone cepstral coefficients (GTCCs) [11], constant Q cepstral coefficients (CQCCs) [12], and acoustic ternary patterns (ATP) [13] are getting popularity for the extraction of the useful feature from the input audio signal. Traditionally, MFCC feature vectors were popular for the feature extraction tasks; however, nowadays, GTCC, ATP, and CQCC-based feature extraction techniques are also getting popularity.

For classification of the input audio samples into spoofed or bonafide audios, initially, the focus of the researchers was toward the use of Gaussian Mixture Model (GMM) [14] based classifier. However, after advancement in artificial intelligence, various machine learning techniques such as random forest (RF) [13], Naïve Bayes (NB) [13], and K-nearest neighbor (KNN) [13], and different deep learning algorithms such as long short-term memory (LSTM) [15], convolutional neural network (CNN) [16], and recurrent neural network (RNN) [17] are becoming popular for speaker verification tasks. For any biometric system, the equal error rate (EER) [18] value is considered as the optimal evaluation criterion to measure the performance of the speaker verification system. EER value generally defines as the rate at which the system is accepting the false values and rejecting the true values. The main focus of the chapter is to analyze the different classification models for the speaker verification tasks. The detailed description of various back-end classification models used in verification of speakers in ASV systems is given in the next section.

The organization of the rest of the chapter is as follows: Section 8.2 discusses the various classification models used to classify audio samples into cloned or original audios in speaker verification task. In Section 8.3, detailed discussion related to the need of validation and types of validation performed in machine and deep learning is done. Section 8.4 explores the various assessment criteria available for evaluating the efficiency of ASV system. Section 8.5 performs the analysis of different classification models that are used by different researchers to improve the performance of ASV system. Section 8.6 includes the conclusion and future scope related to different classification models used in ASV system.

8.2 BACK-END CLASSIFICATION MODELS

When the audio feature vector is fed to classification model, it classifies the audio samples into original human audio or cloned audio sample. First, the classification model is created. After creation of classification model, the hyper parameters are set. Hyper parameters are those parameters that control the learning process. After this, training of the classifier using training feature vector is performed. While training, the validation is also performed. In next step it is checked that whether the results show improvement over previous settings of hyper parameters. If no, then hyper parameters are set again and all the previous steps will run again. If yes, then the testing of model is performed using testing audio feature vector. At last, model gets evaluated using different evaluation criteria and classifies the audio samples into bonafide or spoofed audio samples. Figure 8.2 shows the steps during classification of audio samples. Classification models are divided into two parts. One is machine learning models and other is deep learning models. Different machine learning

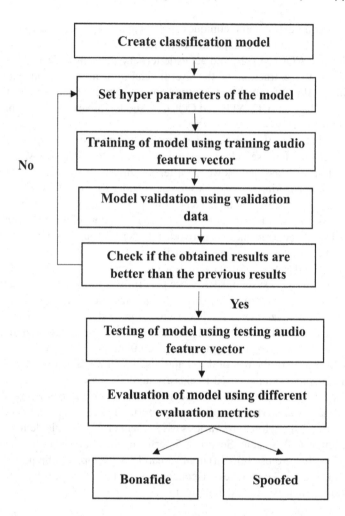

FIGURE 8.2 Steps during the classification of audio samples.

and deep learning classification models such as RF, NB, KNN, LSTM, CNN, and bidirectional LSTM [19,20] are widely used for classification of the audio samples into spoofed and original audio samples. The fundamentals related to different classification models are as follows.

8.2.1 MACHINE LEARNING MODELS

Machine learning models are widely used in classification of text, audios, etc. The brief introduction of popular machine learning algorithms that are used for classification of audio samples is as follows:

- **Naïve Bayes (NB)**

  NB classification algorithm is based on Bayes theorem. It is one of the most common machine learning algorithms that is used for the classification task of audio samples in speech processing domain. Bayes rule calculates the probability of an object by considering the prior knowledge. It depends on the conditional probability. The notation $P(M \mid N)$ means that probability of event M when event N has already been occurred [13,21]. The following equation shows the formula of Bayes rule:

$$P(M \mid N) = \frac{P(N \mid M) * P(M)}{P(N)} \tag{8.1}$$

where $P(M \mid N)$ represents *posterior probability*, $P(M \mid N)$ represents likelihood probability, $P(M)$ denotes prior probability, and $P(N)$ shows marginal probability.

- **K-Nearest Neighbor (KNN)**

  KNN is a supervised learning algorithm for classification and widely used for binary classification. KNN classification algorithm depends on the similarities between the new data and the available data. When the audio feature vector is fed to the KNN model then according to the similarities, the algorithm divides the new data in feature vector into the different classes. In KNN, first, the number of neighbors (N) is selected. And Euclidean distance of all k neighbors is calculated. Then in all categories, the number of data samples is calculated. Whenever new data (feature vector) for the new audio sample comes, on the basis of the Euclidean distance it gets divided into the different categories [13,21]. Euclidean distance between two points A and B having data points (X_1, Y_1) and (X_2, Y_2) respectively, can be calculated as follows:

$$\text{Euclidean distance} = \sqrt{(X_2 - X_1)^2 + (Y_2 - Y_1)^2} \tag{8.2}$$

- **Random Forest**

  RF is based on an ensemble learning algorithm. Ensemble learning is the technique in which output from the different classifiers is taken into consideration to classify the input audio sample into bonafide or spoofed audio sample. In random forest, output from multiple decision trees is considered instead of single decision tree which results in higher accuracy and solves the problem of overfitting which is more prevalent in decision trees [13,22]. The main advantage of RF algorithm over other classifiers is that it can easily handle the large datasets and provides good performance as compared to decision trees. Figure 8.3 shows the random forest classifier for classifying the audio samples for speaker verification task in speech processing.

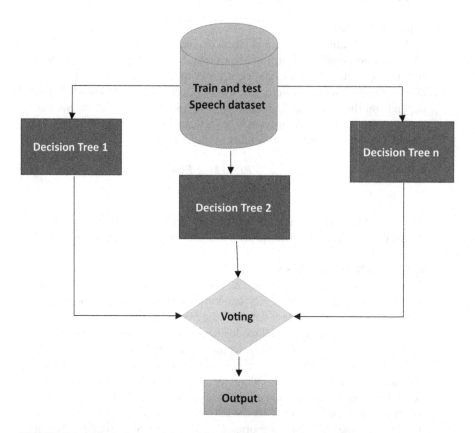

FIGURE 8.3 Block diagram of classification using random forest classifier.

8.2.2 Deep Learning Models

In artificial intelligence, deep learning-based classifiers are getting popularity to classify the audio samples into bonafide (original human audio) or spoofed (cloned) samples. Various popular deep learning-based models used for the classification of audio samples are explained as follows:

- **Convolutional Neural Network (CNN)**

    CNN is one of the most popular deep learning algorithms for classifying images and audio samples. The images and audio samples are considered in matrix form. On these, matrix the operation of mathematics called convolution is performed, which is why the name CNN name is given to this technique. CNN mainly consists of three main layers, i.e., convolutional layer, pooling layer, and fully connected layer [23]. In convolutional layer, the convolution operation is performed between the matrix generated either from the input image or audio sample and the filter having the size $K*K$, i.e., square matrix. The next layer is the pooling layer. The main task of this layer is to reduce the size of the output of the convolutional layer. In simple words, it generally lowers the size of the feature vector. Bias and weights associated with different neurons are considered in a fully connected layer. First, input flattening is performed and then passed to the fully connected layer for the classification task [23,24]. When the data are fed to the fully connected layer, there might be overfitting. So, to prevent the effect of over-fitting, drop-out layers are used. The drop-out layer randomly drops out some neurons to avoid overfitting. Figure 8.4 shows the different layers of the CNN. Different activation functions such as reLU, tanh, SoftMax, and sigmoid are widely used in classification.

- **Recurrent Neural Network (RNN)**

  RNN is the type of neural network in which the output of one layer is fed as the input to another layer. RNN is very useful in the cases where the prediction of the subsequent work of any given sentence needs to be performed. Nowadays, researchers are also using RNN in speech processing tasks. RNN consists of an input layer, hidden layer, and output layer. It provides the same bias and weights to all the layers. Hence, it reduces the increasing parameter complexity and delivers the output of the previously hidden layer to the next layer as input to improve the memorization of the output of each layer. Figure 8.5 shows the block diagram of the RNN.

- **Long Short-Term Memory (LSTM)**

  LSTM model was developed to overcome the disadvantage of RNN. RNN mainly suffers from vanishing gradient problems, i.e., when training the neural network on backpropagation, there is no updating in the value of the weights. The layer of LSTM consists of the connection of memory block in the recurrent fashion. Hence it becomes suitable for sequence pre-diction tasks and also for speech processing. In LSTM network, there are three gates. First is input gate, and other two are output gate and forget gate, respectively. The input gate's job is to evaluate the new information

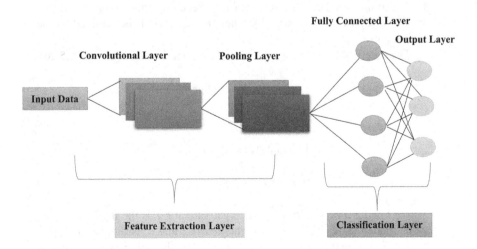

FIGURE 8.4 Layers in convolutional neural network.

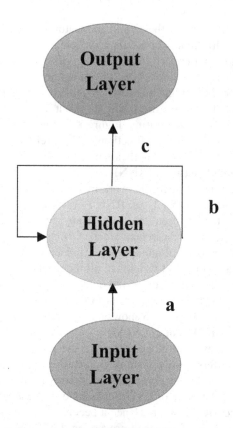

FIGURE 8.5   Block diagram of recurrent neural network.

carried by the input and determine its significance. The forget gate is used to determine whether the data from the preceding time stamp should be kept or dismissed. The value of the next hidden state is determined by the output gate.

The output of each gate is represented by the following equations [25,26]:

$$f_t = \emptyset(Wt_{fg} * |h_{t-1} * i_t| + b_f \tag{8.3}$$

$$i_{tp} = \emptyset(Wt_{ig} * |h_{t-1} * i_t| + b_i \tag{8.4}$$

$$V_i = \tanh(W_c * |h_{t-1} * i_t| + b_c \tag{8.5}$$

$$V_t = f_t * V_{t-1} + i_t * V_i \tag{8.6}$$

where t is the time stamp, f_t is the forget gate at t, i_t is the input, h_{t-1} is the prior hidden state, Wt_{fg} is the weights matrix, and b_f is the bias in equation (8.3). In equations (8.4–8.6), Wt_{ig} represents the weight matrix between the output gate and the input gate, V_i is the result generated after applying the tanh activation function, W_c is the weight matrix between cell state information and the previous output of the network, and V_t is cell state information.

$$o_t = \varnothing(W_O * |h_{t-1} * i_t| + b_o \qquad (8.7)$$

$$Op_t = o_t * \tanh(V_t) \qquad (8.8)$$

where o_t is the output gate at t, W_o and b_o denote the weight and bias of o_t, respectively, and Op_t denotes the final output of LSTM.

- **Gated Recurrent Unit (GRU)**

  The complexity of GRU model is less as compared to LSTM model. It provides faster and better results as compared to LSTM and is widely used when training samples are small in amount [27]. GRU consists of two gates one is reset gate and other one is update gate. The role of update gate is to determine how much past information is to be retained for making the predictions for future. The reset gate helps to determine how much previous information need to forget. The equation for output of update and reset gates is as follows:

$$U(t) = \partial(W_u x_t + U_u h_{t-1}) \qquad (8.9)$$

$$R(t) = \partial(W_r x_t + U_r h_{t-1}) \qquad (8.10)$$

where $U(t)$ and $R(t)$ represent the output of update gate and reset gates, respectively, W_u, U_u, W_r, and U_r show the weights, and x_t is the input, and h_{t-1} is the output of previous $t-1$ states.

8.3 VALIDATION

In artificial intelligence, validation [28] often entails testing the system on the testing dataset. This testing dataset is not a fresh dataset, but rather a subset of the training dataset. Some of the training dataset is put aside for training, while others are placed aside for testing [29]. In order to classify the audio samples into bonafide or spoofed audios, the classification model is trained with different hyper parameters. With the help of validation, the best hyperparameters are selected for the particular model that gives high validation accuracy. Validation testing helps to select the best hyper parameters for the model. The validation techniques also determine the underfitting and overfitting of the classification model. Underfitting is the situation in which the model is neither providing good accuracy on training data nor on testing data. Overfitting occurs when the model generalizes good on the training data but does not show good performance on the testing data. Different validation techniques that are popular to improve the performance of the machine learning and deep learning models are as follows:

8.3.1 HOLD OUT VALIDATION

In holdout cross validation, the training speech feature vector is divided into two parts. One is training speech feature vector part, which will be used for training of the classification model, and another part is testing speech vector part, which will be used for the testing of the classification model [30]. As training is performed on larger dataset, the for-train test split, 80% of data will be reserved for training and remaining 20% of data will be sued for testing. This validation process is the easiest to understand but it has some drawbacks. The major drawback is that the testing data may contain the useful information needed for training of the model and training dataset missed out all the essential information required for training. It may lead to low accuracy during training and validation.

8.3.2 K-FOLD CROSS VALIDATION

In order to overcome the drawbacks of holdout validation, k-fold cross validation is used [31]. In K-fold cross validation, subsets of the training data are formed. K generally defines number of the subsets to be formed. These subsets contain the random data, i.e., subsets may have imbalance classes [29]. It means that one subset may have higher number of bonafide samples as compared to spoofed samples and vice versa. Figure 8.6 shows the block diagram of k-fold cross validation when the number of folds is set to 5.

First, during iteration one, five folds of training data are formed. Four folds all together are reserved for training of the classification model and first subset set is used for testing. Similarly for the second iteration, the second subset will get used for testing of the classification model and remaining four subsets all together will

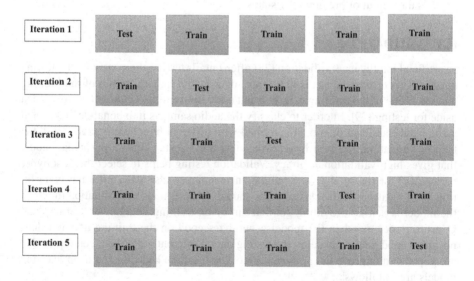

FIGURE 8.6   k-fold cross validation when the number of folds is set to 5.

be used for training of the system. And this process will get repeated till the fifth iteration. It is important to note that only one subset will be reserved for testing at a time, and the remaining subsets all together will be used for training.

8.3.3   STRATIFIED K-FOLD CROSS VALIDATION

In k-fold cross validation, each subset contains the random distribution of the data. It means that one class may have more samples as compared to another class. It may lead to low accuracy. In order to solve this problem, stratified k-fold cross validation is used [32]. In these techniques, between each subset, all the classes are equally distributed. It means that each subset contains a generalized form of the dataset. It may lead to improve the validation accuracy. This is the only difference between k-fold and stratified k-fold cross validation. Rest all the steps in both the techniques are same.

8.3.4   MONTE CARLO CROSS VALIDATION

In Monte Carlo cross validation [33], the training audio feature is divided into train test split. This split can be 70:30, 80:20, and 75:25, i.e., different train test split for all the iterations. Roughly, a large number of iterations such as 500 or 1,000 are used in this process. At the end, average accuracy of all the iterations is taken to calculate the final validation accuracy. This process is the enhancement of holdout validation, as in this approach, different iterations are there and each iteration contains different train test splits. The major disadvantage of this approach is that some of the data points may occur many times in the testing set but some might never occur in the testing set.

8.4   EVALUATION METRICS

Evaluation metrics are the performance evaluation criteria that are used to measure the performance of machine learning and deep learning algorithms. While evaluating the classification model, there can be four possibilities: false positive (FPos), true negative (TNeg), true positive (Tpos), and false negative (FNeg). TPos samples are those that are correctly predicted, FPos samples are those that are mistakenly forecasted, TNeg samples are those that are correctly predicted, and FNeg samples are those that are incorrectly predicted. The brief introduction to different evaluation criteria that are used to measure the performance of the proposed system is as follows.

8.4.1   EQUAL ERROR RATE (EER)

EER value is one of the superlative evaluation criteria for the evaluation of the ASV system [34]. It generally outlines the proportion at which system discards the bonafide audio data, i.e., false rejection rate (FRR), and admits the spoofed audio data, i.e., FRR. EER usually sets the threshold value of FRR and FAR. The point at which FRR and FAR becomes equals that point is entitled as the ERR value. The formula for FAR and FRR is as follows:

$$FAR = \frac{\text{Number of False Acceptance possibilties}}{\text{Total number of the utterances}} \qquad (8.9)$$

$$FRR = \frac{\text{Number of False Rejection possibilties}}{\text{Total number of the utterances}} \qquad (8.10)$$

8.4.2 ACCURACY

Accuracy is another evaluation criterion that is used to measure the performance of classification model. Accuracy is the measure of number of the samples that model correctly predicts out of the total number of the samples [34]:

$$Accuracy = \frac{\text{Number of correctly predicted samples}}{\text{Total number of the samples}} \qquad (8.11)$$

$$Accuracy = \frac{\text{TPos+TNeg}}{\text{TPos+ FPos+TNeg+FNeg}} \qquad (8.12)$$

8.4.3 PRECISION (PRE)

Precision is the fraction of number of the audio samples that are correctly predicted, i.e., Tpos to the total number of the correctly predicted and incorrectly predicted samples of positive class i.e., Tpos and FPos [34].

$$Precision = \frac{\text{TPos}}{\text{TPos} + \text{FPos}} \qquad (8.13)$$

8.4.4 RECALL (RE)

Recall basically defines out of all the predicted samples of one class how many of the samples are correctly predicted by our classification model. Recall is the fraction of number of TPos to TPos and FNeg [34].

$$Recall = \frac{\text{TPos}}{\text{TPos} + \text{FNeg}} \qquad (8.14)$$

8.4.5 AREA UNDER CURVE (AUC)

Area under ROC curve in machine learning is the performance evaluation criteria that is used to measure the performance of the classification algorithm. It defines the rate at which our classification algorithm is able to find differences between the binary classes of the samples. More the AUC value better our system classification would be and vice versa. The ROC curve is plotted between the Tpos rate (recall) and FPos rate [35]. Formula for FPos rate is as follows:

$$\text{False Positive Rate} = \frac{FPos}{TNeg + FPos} \quad (8.15)$$

8.4.6 F1 SCORE

F1 score is the evaluation criterion widely used in the binary classification problem. It is the harmonic mean between the precision and recall values. In order words, we can say it is the weighted average between recall value and precision value. It has actual application when our dataset contains the irregular distribution of the classes [34]. The formula for the calculation of F1 score is as follows:

$$\text{F1 Score} = \frac{2 * Pre * Re}{Pre + Re} \quad (8.16)$$

8.4.7 MEAN AVERAGE PRECISION (AP)

AP is another popular evaluation metric that is used for measuring efficiency of the classification algorithm. It is defined as taking the mean of precisions over all the data sample groups reliant on diverse trials that happen [36]. The formula for the calculation of AP is as follows:

$$AP = \frac{\sum_{i=1}^{n} \text{Precision of given class}}{\text{Total number of samples of particular class}} \quad (8.17)$$

8.5 ANALYSIS OF DIFFERENT CLASSIFICATION MODELS

The work of Sadhu et al. [34] mainly focused on the hybrid of LSTM and CNN model at back end to classify the audio samples into bonafide or spoofed audios. First, the authors have extracted Mel spectrograms from the audio samples and these spectrograms are fed to CNN classification model. From the same audio samples, CQCC features are extracted and fed to LSTM based classification model. The output of both the models gets combined to provide the final output as bonafide or spoofed audio. The proposed system by them provides 3.6% EER in 1PR attacks and 2.96% EER in 2PR attacks, respectively.

The work of Mittal and Dua involves different deep learning models and combination of different deep learning models to classify the audio samples into bonafide or spoofed audios. First, the authors [37] have proposed CQCC–2DCNN (two-dimensional convolutional neural network)-based ASV system. For extraction of CQCC features the authors have used 30 static + 30 first order + 30 second order features. Their first proposed system provides 5.5% EER, 0.101 min t-DCF value against logical access (LA) attacks, and 6.2% EER and 0.122 min t-DCF value against PA attacks. Second, the authors [37] have proposed CQCC–LSTM-based system. The proposed system provides 0.42% EER against LA attacks and 0.51% EER value against PA attacks. In the third proposed system [12], the authors have combined different deep learning models such as LSTM with time-distributed wrappers and CNN.

The authors conclude that hybrid deep learning models improve the classification performance of ASV system.

Malik et al. [13] proposed ATP+GTCC-based feature extraction technique. The extracted feature vectors are fed to different classification models such as Error-Correcting Output Coding-Support Vector Machine (ECOC-SVM), bagged trees, decision trees, Naïve Bayes, KNN, and bidirectional LSTM. The authors conclude that hybrid feature extraction at front end and machine learning model at back end improves the efficiency of ASV system. The proposed system by them provides 0.6% EER against 1PR and 2PR attacks and 1% EER value under PA attacks.

Li et al. [38] proposed attention-based LSTM to classify the audio samples into bonafide or cloned audio samples. The authors conclude that the attention-based LSTM provides nearly 13.81% improvement in EER value as compared to LSTM.

Faisal et al. [39] used MFCC feature extraction to extract the important features from the audio samples. After that, the extracted feature vectors are given to DNN and GMM classifiers individually. The authors conclude that the DNN-based classifier provides better results (18.1% EER) as compared to GMM-based classifier (19% EER).

Dua et al. [40] have proposed a hybrid of deep neural networks such as LSTM with time-distributed dense layers, and spatial convolution (SC) and temporal convolution (TC)-based DNN to perform the classification of the audio samples into bonafide or spoofed audios. The hybrid deep neural network model provides 0.6% EER against both LA and physical access (PA) (ASVspoof 2019 dataset) attacks.

Kumar and Bharathi [41] proposed FBCC feature extraction technique to extract the useful information from the audio samples. The extracted audio feature vector is fed to GMM-based classifier. The proposed system provides 0.16 min t-DCF value against LA attacks and 0.25 min t-DCF value against PA attacks. Table 8.1 provides the performance analysis of ASV systems that use different feature extraction and classifiers to classify the audio samples into original human audio or spoofed audio.

Further, from the performance analysis of different classifiers, it can be concluded that the classification model that performs the validation of the obtained results using different validation techniques, that classifier provides better performance as compared to others.

8.6 CONCLUSION

For industry automation, voice technologies are very helpful to automate the production process. Speaker verification performs the classification of voice samples into bonafide or spoofed audios. The back-end classification model plays a significant role to improve the accuracy of ASV system. Traditionally, machine learning-based models were popular for spoofing detection task in speaker verification system. However, today deep learning models and hybrid of various deep learning models are getting researchers' attention. Although the hybrid deep learning models at back end improve the performance of ASV system, it also increases the overall complexity of the system. Further to increase the performance of classification model to classify the audio samples into bonafide or spoofed, the validation procedure can be applied. For evaluating the ASV system, along with EER value different performance evaluation

TABLE 8.1

Analysis of Different Classification Models Used at Back End of ASV System

Work	Front End	Back End	Validation	Results	Remarks
Sadhu et al. [34]	Mel spectrograms, CQCC	LSTM, CNN	K-fold cross validation	3.6% EER (1PR), 2.96% EER (2PR), 97.6% Accuracy (1PR), 97.78% Accuracy (2PR)	Hybrid deep learning model of LSTM+CNN at back end improves the performance of overall system as compared to system that uses standalone CNN and LSTM models.
Mittal and Dua [12]	Static dynamic CQCC features	LSTM with time-distributed wrappers, CNN, Voting protocol	Validation using development set	0.9% EER	Hybrid deep learning models improve the performance of ASV system.
Dua et al. [40]	MFCC, IMFCC, CQCC	LSTM with time-distributed dense layers, SC and TC-based DNN, Ensemble model of all these models	—	0.6% EER using ensemble model	CQCC features with ensemble deep neural network provide better EER as compared to other models and existing systems.
Mittal and Dua [37]	30 static + 30 first order + 30 second order CQCC features	2D CNN	—	5.5% EER and 0.101 t-DCF for LA attacks, 6.2% EER and 0.122 t-DCF for PA attacks	Combination of 2D CNN with CQCC features increases the efficiency of ASV system.
Malik et al. [13]	Hybrid ATP+GTCC features	ECOC-SVM, Decision Trees, KNN, NB, Bagged Trees, Bidirectional LSTM	K-fold Cross Validation	0.6% EER (VSDC), 1% EER (PA)	When hybrid feature extraction technique is used at front end, machine learning algorithms show great performance as compared to deep learning techniques.
Li et al. [38]	CQCC features	Attention-based Long Short-Term Memory	—	20.32% EER (Traditional LSTM), 16.86% EER (Attention-based LSTM)	Attention-based LSTM provides better results as compared to other models.
Mittal and Dua [15]	CQCC	LSTM classification model	—	0.42% EER (LA attacks), 0.51% EER (PA attacks)	LSTM model outshines other baseline models when CQCC features are fed to it.
Faisal et al. [39]	MFCC	DNN, GMM	—	18.1% EER (DNN), 19.0% EER (GMM)	Deep neural networks provide good results as compared to GMM.
Kumar and Bharathi [41]	FBCC	GMM classifier	—	0.16 t-DCF (LA attacks), 0.25 t-DCF (PA attacks)	GMM-based classifier improves the min t-DCF value when FBCC features are fed to it.

criteria such as F1 score, precision, recall, and accuracy can be used. Further in future scope to increase the performance and to decrease the overall complexity of ASV system, the ASV system with hybrid feature extraction at front end and less complex classification model such as light convolutional neural network (LCNN) can be used.

REFERENCES

1. Z. Wu, N. Evans, T. Kinnunen, J. Yamagishi, F. Alegre, and H. Li, "Spoofing and countermeasures for speaker verification: A survey," *Speech Commun.*, vol. 66, pp. 130–153, 2015.
2. A. Indumathi and E. Chandra, "Survey on speech synthesis," *Signal Process. An Int. J.*, vol. 6, no. 5, p. 140, 2012.
3. A. Kain and M. W. Macon, "Design and evaluation of a voice conversion algorithm based on spectral envelope mapping and residual prediction," in *2001 IEEE International Conference on Acoustics, Speech, and Signal Processing. Proceedings (Cat. No. 01CH37221)*, 2001, vol. 2, pp. 813–816.
4. M. Liu, L. Wang, Z. Oo, J. Dang, D. Li, and S. Nakagawa, "Replay attacks detection using phase and magnitude features with various frequency resolutions," in *2018 11th International Symposium on Chinese Spoken Language Processing (ISCSLP)*, 2018, pp. 329–333.
5. V. Vestman, T. Kinnunen, R. G. Hautamäki, and M. Sahidullah, "Voice mimicry attacks assisted by automatic speaker verification," *Comput. Speech Lang.*, vol. 59, pp. 36–54, 2020.
6. Z. Wu et al., "ASVspoof 2015: the first automatic speaker verification spoofing and countermeasures challenge," 2015.
7. H. Delgado et al., "ASVspoof 2017 Version 2.0: meta-data analysis and baseline enhancements," 2018.
8. Y. Yang et al., "The SJTU robust anti-spoofing system for the ASVspoof 2019 challenge," in *Interspeech*, 2019, pp. 1038–1042.
9. R. Baumann, K. M. Malik, A. Javed, A. Ball, B. Kujawa, and H. Malik, "Voice spoofing detection corpus for single and multi-order audio replays," *Comput. Speech Lang.*, vol. 65, p. 101132, 2021.
10. S. Nakagawa, L. Wang, and S. Ohtsuka, "Speaker identification and verification by combining MFCC and phase information," *IEEE Trans. Audio. Speech. Lang. Processing*, vol. 20, no. 4, pp. 1085–1095, 2011.
11. M. Dua, R. K. Aggarwal, and M. Biswas, "GFCC based discriminatively trained noise robust continuous ASR system for Hindi language," *J. Ambient Intell. Humaniz. Comput.*, vol. 10, no. 6, pp. 2301–2314, 2019.
12. A. Mittal and M. Dua, "Static–dynamic features and hybrid deep learning models based spoof detection system for ASV," *Complex Intell. Syst.*, vol. 8, pp. 1–14, 2021.
13. K. M. Malik, A. Javed, H. Malik, and A. Irtaza, "A light-weight replay detection framework for voice controlled IoT devices," *IEEE J. Sel. Top. Signal Process.*, vol. 14, no. 5, pp. 982–996, 2020.
14. A. Kuamr, M. Dua, and T. Choudhary, "Continuous Hindi speech recognition using Gaussian mixture HMM," in *2014 IEEE Students' Conference on Electrical, Electronics and Computer Science*, 2014, pp. 1–5.
15. A. Mittal and M. Dua, "Constant Q cepstral coefficients and long short-term memory model-based automatic speaker verification system," in *Proceedings of International Conference on Intelligent Computing, Information and Control Systems*, 2021, pp. 895–904.

16. B. Chettri, S. Mishra, B. L. Sturm, and E. Benetos, "Analysing the predictions of a cnn-based replay spoofing detection system," in *2018 IEEE Spoken Language Technology Workshop (SLT)*, 2018, pp. 92–97.
17. T. Tan et al., "Speaker-aware training of LSTM-RNNs for acoustic modelling," in *2016 IEEE International Conference on Acoustics, Speech and Signal Processing (ICASSP)*, 2016, pp. 5280–5284.
18. J.-M. Cheng and H.-C. Wang, "A method of estimating the equal error rate for automatic speaker verification," in *2004 International Symposium on Chinese Spoken Language Processing*, 2004, pp. 285–288.
19. S. Duraibi, W. Alhamdani, and F. T. Sheldon, "Replay spoof attack detection using deep neural networks for classification," in *2020 International Conference on Computational Science and Computational Intelligence (CSCI)*, 2020, pp. 170–174.
20. C. Hanilci, T. Kinnunen, M. Sahidullah, and A. Sizov, "Spoofing detection goes noisy: An analysis of synthetic speech detection in the presence of additive noise," *Speech Commun.*, vol. 85, pp. 83–97, 2016.
21. T. G. Dietterich, "Ensemble methods in machine learning," in *International Workshop on Multiple Classifier Systems*, 2000, pp. 1–15.
22. G. Biau and E. Scornet, "A random forest guided tour," *Test*, vol. 25, no. 2, pp. 197–227, 2016.
23. S. Albawi, T. A. Mohammed, and S. Al-Zawi, "Understanding of a convolutional neural network," in *2017 International Conference on Engineering and Technology (ICET)*, 2017, pp. 1–6.
24. A. Krizhevsky, I. Sutskever, and G. E. Hinton, "Imagenet classification with deep convolutional neural networks," *Adv. Neural Inf. Process. Syst.*, vol. 25, 2012, pp. 1097–1105.
25. K. Greff, R. K. Srivastava, J. Koutník, B. R. Steunebrink, and J. Schmidhuber, "LSTM: A search space odyssey," *IEEE Trans. Neural Networks Learn. Syst.*, vol. 28, no. 10, pp. 2222–2232, 2016.
26. A. Graves and J. Schmidhuber, "Framewise phoneme classification with bidirectional LSTM and other neural network architectures," *Neural Networks*, vol. 18, no. 5–6, pp. 602–610, 2005.
27. R. Dey and F. M. Salem, "Gate-variants of gated recurrent unit (GRU) neural networks," in *2017 IEEE 60th International Midwest Symposium on Circuits and Systems (MWSCAS)*, 2017, pp. 1597–1600.
28. P. Refaeilzadeh, L. Tang, and H. Liu, "Cross-validation," *Encycl. Database Syst.*, vol. 5, pp. 532–538, 2009.
29. T. Fushiki, "Estimation of prediction error by using K-fold cross-validation," *Stat. Comput.*, vol. 21, no. 2, pp. 137–146, 2011.
30. S. Yadav and S. Shukla, "Analysis of k-fold cross-validation over hold-out validation on colossal datasets for quality classification," in *2016 IEEE 6th International Conference on Advanced Computing (IACC)*, 2016, pp. 78–83.
31. J. Wieczorek, C. Guerin, and T. McMahon, "K-fold cross-validation for complex sample surveys," *Stat*, vol. 11, p. e454, 2022.
32. S. Alrabaee, "A stratified approach to function fingerprinting in program binaries using diverse features," *Expert Syst. Appl.*, vol. 193, p. 116384, 2022.
33. P. Smyth, "Clustering using Monte Carlo cross-validation," *Kdd*, vol. 1, pp. 26–133, 1996.
34. M. Dua, A. Sadhu, A. Jindal, and R. Mehta, "A hybrid noise robust model for multireplay attack detection in automatic speaker verification systems," *Biomed. Signal Process. Control*, vol. 74, p. 103517, 2022.
35. A. P. Bradley, "The use of the area under the ROC curve in the evaluation of machine learning algorithms," *Pattern Recognit.*, vol. 30, no. 7, pp. 1145–1159, 1997.

36. S. Ding, Q. Wang, S. Chang, L. Wan, and I. L. Moreno, "Personal VAD: Speaker-conditioned voice activity detection," *arXiv Prepr.* arXiv1908.04284, 2019.

37. A. Mittal and M. Dua, "Automatic speaker verification system using three dimensional static and contextual variation-based features with two dimensional convolutional neural network," *Int. J. Swarm Intell.*, vol. 6, no. 2, pp. 143–153, 2021.

38. J. Li, X. Zhang, M. Sun, X. Zou, and C. Zheng, "Attention-based LSTM algorithm for audio replay detection in noisy environments," *Appl. Sci.*, vol. 9, no. 8, p. 1539, 2019.

39. M. Y. Faisal and S. Suyanto, "SpecAugment impact on automatic speaker verification system," in *2019 International Seminar on Research of Information Technology and Intelligent Systems (ISRITI)*, 2019, pp. 305–308.

40. M. Dua, C. Jain, and S. Kumar, "LSTM and CNN based ensemble approach for spoof detection task in automatic speaker verification systems," *J. Ambient Intell. Humaniz. Comput.*, vol. 13, pp. 1–16, 2021.

41. S. Rupesh Kumar and B. Bharathi, "A novel approach towards generalization of countermeasure for spoofing attack on ASV systems," *Circuits, Syst. Signal Process.*, vol. 40, no. 2, pp. 872–889, 2021.

9 Trending IoT Platforms on Middleware Layer

Neha Katiyar and Priti Kumari
Noida Institute of Engineering & Technology

Surabhi Sakhshi
Indian Institute of Management

Jyoti Srivastava
Madan Mohan Malviya University of Technology

CONTENTS

9.1   INTRODUCTION

Internet of Things (IoT) is the word that was first introduced to this world by Kevin Mshton in 1999. IoT means 'Internet' is the network of networks, and connectivity is needed to connect the things autonomously. It is expected that by 2020 around 60 million devices may be connected to the Internet. The IoT architecture consists of different layers in it. The sensors and actuators layers, transit and communication layer, middleware layer, and application layer. Middleware act as a bridge between the network and applications. It is linked between the different software and is hidden between the OS and application layers. The application layer includes different types of applications that can run on various IoT platforms. The basic principle IoT follows is as we give commands by speaking, the device performs the same task as

DOI: 10.1201/9781003321149-9

we give to them. It can be achieved by various technologies such as Bluetooth, WiFi (wireless fidelity), Zigbee, and Z-Wave. They constantly maintained undisputed proof of the progress of IoT platforms achieving that technical revolution.

IoT platforms enable interoperability for heterogeneous devices and technologies. There are so many platforms for IoT that can be efficiently designed for the literature and industry, mainly healthcare fields for monitoring and energy conservation and psychological fields for child monitoring. In middleware layer solutions, programmers must know about the new software specifications because they are implemented on the latest software packages. These tasks consumed a lot of time. In IoT, different corporate persons and industrialists will use various software that might be incompatible. Billion devices are connected with IoT (sensors, printers, mobiles, laptops, cameras, and projectors), and their heterogeneous IoT platforms can perform essential tasks. Prominent technology companies created their own IoT platform, the IoT activity extensive framework. The top companies have strong competencies and their name mentioned are Amazon web services, Ayla Agile IoT platform, Azure IoT suite, Bosch IoT suite, Jasper control center, Oracle IoT cloud services, GE Predix, SAP cloud platform, PTC Thing work, and Watson IoT platform.

As the IoT, as mentioned above, platforms nowadays cover various sectors. IoT platforms have gained more importance in the last decades. The criteria of these platforms' functionality, strategic techniques, easy availability, and cloud connectivity will increase the platform's position in the market. The word IoT consists of the word Internet. Due to this reason, the entire thing that works on the Internet currently uses significantly lighter IoT versions. Big data analysis is the most frequent wireless method for connecting the Internet. It performs such that the Internet protocol stack does not apply any restrictions on IoT in the account. Bluetooth 5, Zigbee, and WiFi are these technologies used for better connectivity with the Internet in IoT. Bluetooth 5.0 is the trending variant used for communication between various intelligent homes, IoT platforms, and IoT devices. (Bluetooth range is limited; it supports only 5 m in outdoor environments, and 200 m and 40 m indoors).

Perceiving the decisive role of platforms in IoT platforms, we focused on that paper as IoT middleware platforms or an instance model of IoT architecture. This chapter's proposal of IoT security features in IoT middleware is mainly designed for a protected environment for IoT. This chapter also highlights the significance of IoT middleware and platforms and their essential role with 6G network connecting technology. The remaining piece consists of the following criteria: Section 9.2 provides an analysis of security in IoT middleware, IoT networks, and related work. Section 9.3 IoT platforms. Section 9.4 presents the categories of IoT platforms for industry. Section 9.5 presents the trending IoT platform for Industry Revolution 4.0. Section 9.6 proposes a technique for IoT platforms connectivity with the 6G network. Finally, Section 9.7 is the conclusion of the chapter.

9.2 RELATED WORK

In general, based on various literature research smart network connectivity, we observed that IoT platforms are categorized into two sections:

1. Cloud-based systems with IoT devices.
2. Device gateways are directly connected to the cloud through networks.

In 2020, Viswanathan et al. [13] focused on the 6G network is gradually taking momentum. This chapter discusses different interface connectivity with 6G networks such as man–machine interfaces, ubiquitous interfaces, multilevel sensor data, or IoT platform connectivity. The article also shows the foundation of the 6G interface air network and its connectivity with different devices. It is the most beneficial networking technology for connecting IoT devices with the middleware layer platform. Nevertheless, a 6G network is undoubtedly taking the shape of IoT platforms.

In 2021, Barakat et al. [14] also highlighted the 6G connecting technology with IoT platforms. 6G will continue to provide its multiple feature network services for all digital devices. Smart cities will have the future of flying taxis already operated in limited areas in various cities such as Dubai. The regulation, as well as connectivity requirements of flying taxis, required a fully connected IoT platform.

In 2021, Nkenyereye and Hwang [17] mainly highlighted the specialization of IoT systems. They focused on the visual network functions of the IoT platform on different layers. They give an idea about IoT platform communication infrastructure method with network slicing involving 5G and 6G techniques. They designed an algorithm for network slicing of IoT platforms. The algorithm name is virtual IoT slice services functions (VIoT-SSFs). This algorithm improves the network function of IoT platforms. The IoT ecosystem consists of a stratified architecture model and uses this representation to provide more imaginative solutions.

In 2018, da Cruz et al. [1] discussed the recommended dummy for the IoT middleware. It shows designate the correct platform for a specific task and the differentiate between the better and worse IoT solutions because it is a long-term process to make the platform user-friendly. They discussed the features of the IoT platform and made them more compatible without any security point drawbacks. IoT middleware platforms are facing many challenges in providing security.

In 2015, Fersi [2] discussed and classified the challenge facing middleware IoT platforms. They organized the middleware platform approach depending on the generic techniques. They highlighted and explored the significant issues of security, privacy, configuration, and bootstrapping maintenance issues.

Several algorithms are designed to maintain IoT platforms' security and network connectivity. In 2016, Pallavi et al. [3] discussed the security algorithm that protects the IoT data on the middleware layer. The security in middleware is done by applying different algorithms in the fog layer to secure the information it will send to the cloud. In 2021, Nawaz et al. [15] focused on various techniques of IoT platforms. A secure IoT system is made when we apply several transfer protocols and use the most secure connecting technology. 6G is the fastest and connecting technology. In 2018, Kokkonis et al. [4] proposed that several IoT middleware protocols have been used to transfer the data to the cloud. The prominent representatives of IoT are the COAP (constrained application protocol), SOAP (simple object access protocol), MQTT (MQ telemetry transport), and HTTP (hypertext transfer protocol). It concerns the various protocol techniques such as delay, packet loss, and processing of IoT devices. In addition, the categorization of an advanced security layer in middleware platforms that may generate additional features overhead, depending on the restraint of the environmental resources, and these costs cannot be confirmed without security measures. A secure network is formed on these platforms.

In 2019, Silva et al. [5] implemented management for devices and networks in IoT (M4DN.IoT) – a network and device management platform for the IoT. M4DN. IoT network provides a solution for automatic IoT network management and a user compatibility interface that includes information about the network devices. This platform may be used in any smart device (desktop, laptop, and smart meter) and provide access to any location. In 2021, Nkenyereye et al. [17] focused on the working of the IoT platform. The IoT platforms are providing good quality services. Interactions have been bounded due to heterogeneous creation of devices, communication protocols, data types, data rates, quality of service (QoS) requirements, and trust measures between applications. It is also used for maintaining the security of the home environment. Sandor et al. [6] proposed the operational requirements of protection for an IoT platform called security-enhanced interoperability middleware for the IoT.

The proposed IoT platform is named ghost – safeguarding home IoT environments with personalized real-time risk control. IoT devices provide that much comfort to the industries so industries can adapt quickly, and it designed the Industrial IoT (IIoT) procedure. In this IIoT procedure, in 2018, Bakhshi et al. [7] discussed the IIoT and investigated the security and concern of IoT. For the data accumulation and abstraction levels of IIoT, they chose Cisco and Microsoft Azure IoT because the two layers of the Cisco reference architecture model have been selected to determine the various security measures. IIoT is used in industries to develop the best quality products. The establishment of IIoT has provided a new platform designed in the middleware layer called IOP (Internet of production). In 2018, Jorg Hoffman et al. [8] proposed the method of Internet of production for the IoT platforms. The IoT platforms in the merchandise are not fully customized to meet the particular requirements of users and organizations – specifically in the manufacturing industry; to manage these issues, the Internet of production is needed. Additionally, IoT creates a paradigm for a required set of features that represents an architecture of reference models. In designing the IoT middleware platforms and models, many challenges nowadays occur. Some researchers also focused on the challenges of IoT platforms. In 2018, Yadav et al. [9] discussed IoT and its challenges and issues in Indian perspective. It familiarizes to show the growth of IoT in India, its risk factors, and security issues challenge an Indian view. The most critical challenge is maintaining the security of IoT platforms. In 2019, Jin-Young Yu et al. [10] analyzed IoT platform security. Different settled IoT platforms are customized based on media on which similarity is underway, such as one Machine-to-Machine (M2M) and Open Connectivity Foundation (OCF). Different settled IoT platforms are customized based on media on which similarity is underway, such as one M2M and OCF for the analysis and comparison of security elements of national and international IoT platforms.

In 2017, Mario Frustaci et al. [11] proposed a new paradigm in SIOT (social Internet of things), which merged with socialization and allowed humans to connect with devices and share information. They study the following three significant layers: (i) perception layer, (ii) transportation layer, and (iii) application layer. The most unsafe level of the IoT system is the perception layer due to the physical disclosure of the IoT. The IoT middleware layer platforms are also used in the healthcare systems for sending and detecting data of patients to forensics. Macro Conti et al. [12] introduced the significant security and forensics issues in IoT with their potentially promising solutions. They introduced special security, privacy, and protection challenges in IoT environments. In 2021, Jang et al. [16] focused on nonintrusive load monitoring techniques. These techniques are beneficial and efficient for intelligent energy systems. In this technique, they designed IoT system architecture that consists of an application layer, perception layer, appliances layer, communication layer, and middleware layer. They compare this generation technique with ILM (intrusive load monitoring technique) because the energy consumption is more than the NILM (nonintrusive load monitoring) technique. Moreover, the IoT middleware layer provides a sustainable, innovative system that consumes low power. The computational requirement of the middleware layer is very high. In 2020, Swamy et al. [18] focused on cloud, fog, and edge computing connectivity with IoT platforms. In this paper, they summarized the IoT platform and IoT application architecture. They also focused on real-time-based IoT systems. It uses various computing techniques. The middleware layer is the abstract layer between the user interface and deployed devices.

9.3   IoT PLATFORMS

At a high level, an IoT platform supports software that interconnects each hardware, access point, and data network to additional quality of the value chain. An IoT platform is often referred to as a middleware solution, which is the plumbing of the IoT. These platforms are a package that integrates devices and applications, as shown in Figure 9.1. Middleware is the link between different software and applications. The middleware system integrates the distributed application system with heterogeneous network environments. The middleware layer connectivity with hardware and applications is shown in Figure 9.2. The layer is hidden between the OS and application layers. It is connected to all types of devices. The kinds of middleware layer platforms are formed, and their specifications given are as follows:

1. **Message-Oriented Middleware:** this type of middleware software allows additional software components to send and receive messages. This software also provides connectivity to the front and back end of the procedure.
2. **Object Middleware:** it is a dissent masterminded system that sends inquiries about the request of organizations.
3. **Database Middleware:** this middleware layer can be used to communicate with database storage. The data middleware layer can combine if the user wants to extract and retrieve various information from the database.
4. **Exchange Middleware:** the trades exchanging with applications and screens are used in the durability of middleware. It covers the web-application servers.
5. **Implanted Middleware:** this type of middleware layer is considered resemblance and coordination associated with programming and interfaces.

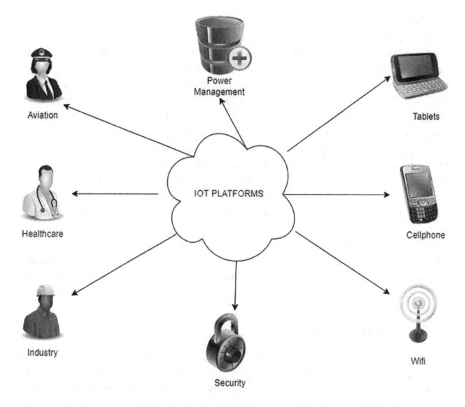

FIGURE 9.1   IoT platform and its applications.

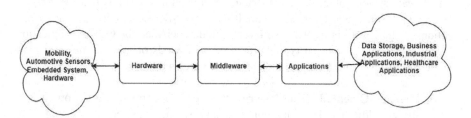

FIGURE 9.2   IoT middleware.

The platform hides the complex infrastructure and supports the enabled IoT solutions. IoT platform increases the market value rate of the IoT sector. These platforms are also called players of the market. The IoT platform market continues to provide strong marketing businesses for IoT data-driven companies. The IoT platform market period from 2015 to 2021. The Club and Group Activity Report (CGAR)'s annual revenue was reaching $1.6 billion in 2021. IoT platform has some functional as well as non-functional requirements in various industries. Functional requirements in the industry such as healthcare and cloud storage.

9.3.1 FUNCTIONAL REQUIREMENTS

Functional requirement is the declaration of the intended operations of a system and its parts. A plan should be designed in the form of functional requirements. Functional requirements can be an instruction set for a device that gives you an idea about assembling a device, which are explained as follows:

- **Diverse Connections:** Ethernet is needed to connect the Internet in the IoT. IoT devices can use various technologies for connectivity over wide applications. The connectivity of IoT supports, various ways of connectivity wired and wireless. Wireless options are Bluetooth, RFID tags, GPRS devices, WLAN, Zigbee, and Z-Wave.
- **Leverage Applications:** when you enter the home, the door is automatically open, lights are on automatically, and AC depends upon automation capabilities that make IoT solutions valuable. These applications provide efficiency with application to the IoT.
- **Industry Range Management**: the number of devices connected in this world of IoT is 28–50 billion. According to the area of devices, the range is for monitoring. IoT sensors can collect the data from where the sensors were placed, such as temperature and moisture level. Actuators perform specific tasks, and it makes things turn on and off. In addition, IoT has different kinds of wearable devices, such as a health-tracking bracelet that can monitor your health positioning at your location. The IoT platform can manage a range of heterogeneous sets of machines.
- **Massive Data:** as we know, about billions of devices are connected with IoT. They will also report the tasks that they performed. Their connection with the IoT platform and with different devices. They will have to transmit each detail of action to one estimation of forces in 2018. IoT generates around 400 terabytes of data. IoT platforms may be able to store and support massive amounts of data.
- **Powerful Analytics:** the massive amount of data stored by IoT platforms. We require powerful analytics tools to recognize which data is helpful for consumers. A compatible platform analytics solution will translate significant amounts of data into valuable and actionable insights.

9.3.2 NONFUNCTIONAL REQUIREMENTS OF IoT PLATFORMS

Nonfunctional requirements substantially show how the system should work and that it depends on the system's behavior. The nonfunctional requirement consists of all those left requirements detained by functional requirements, which is explained as follows:

- **Security:** the most compelling concern in each platform and application is security. In IoT, it is essential to avoid and perform against all sorts of attacks, such as brute-force and Dos, or not to disclose sensitive information, such as user locations, photos, documents, and regular schedules. The significance of data being exposed without any limit platform should do the best protection of user data while providing the mechanism of intrusion detection. Some IoT security solutions are Blackberry, Cisco, Subex, CUJO F-Secure, ZingBox, Luma, etc. These platforms offer better security services to the users.

- **Scalability:** connectivity of billions of devices and Zettabytes of data is discussed, so scalability is the explicit requirement of the IoT platform. IoT platforms should become scalable to grow the network exponentially. Many IoT solutions can achieve their potential only at a scale.
- **Availability:** the reason is that the Internet is needed for connecting IoT devices. The availability of the Internet makes devices work efficiently. Due to the IoT devices being connected to the real-world impact, the IoT platform must provide the highest availability.
- **Maintainability:** the main reason for applying IoT to manage your predictive maintenance. Maintainability means maintaining the performance of IoT devices, compatibility, and maintenance of overall platforms. In this maintenance, more and more systems shipping connectivity is needed.
- **Serviceability:** IoT devices and platforms can work together, and their speed should be maintained. Serviceability means diagnosing problems that should help a rise in the IoT platform while working or providing better services to the users, so users are tangled with media and techniques.
- **Usability:** this means how an IoT platform should be used. The IoT platform must be utilized. It should be used by users. It may not be complex for such technical and non-technical users to operate efficiently.
- **Data Integrity:** we mentioned that terabytes of data should be transferred over to the cloud and data used by the devices. The data are movable to protect information from modification while on its journey from the sensor to the cloud application. Data integrity is one of the most important key aspects of implementing security policies.

9.4 CATEGORIES OF IoT MANAGEMENT PLATFORM FOR INDUSTRIES

The best IoT platforms support functional as well as nonfunctional requirements. The three main categories of IoT industry management platforms are as follows:

1. Device management platforms.
2. Application management platforms.
3. Application development platforms.

9.4.1 DEVICE MANAGEMENT PLATFORMS

One of the topmost concerns of the management of IoT device management is an end-to-end device security provider. IoT devises management includes IoT security. The management of IoT platforms encompasses device positioning administration and monitoring and diagnostics in various industries. IoT creates management functions consisting of enrollment or provisioning association, configuration software updates, and overall management and control in industries. The device management platform provides class-leading lifecycle management for industry development.

9.4.2 IoT Application Management Platforms

IoT Application Enablement Platform (AEP) as a technology essentially offers raise to best of breed, and industry extensible middleware core builds up a set of independent and interconnected solutions for the customers. IoT application enablement: a comprehensive set of devices and enterprise backend should be connected by Software Development Kit (SDKs) and Application Programming Interface (APIs) and the location of well-defined documents and developer resources for IT and electrical industries.

9.4.3 Application Development Platforms

The IoT application development method is a compound of a large-scale technological modernization process. The IoT confirms that it is an extension of interaction between applications and people through new dimensions of "things" for communication. The main principle for developing the IoT application is to ensure the safe collection of data. At the early stage of IoT deployment, deriving domain-specific applications is the primary development strategy for the healthcare industry. In the developing phase, control and manage the high performance of data streaming. IoT platforms connect with the cloud and increase data management efficiency in various healthcare industries.

9.5 TRENDING IoT PLATFORM FOR INDUSTRY REVOLUTION 4.0

IoT platforms set up a new era of cloud computing. IoT middleware solutions are available in the market to increase compatibility worldwide. There are open and closed source middleware platforms given, and the topmost IoT platforms for 2021 are as follows:

1. **Google Cloud Platform:** it provides multifunction secured infrastructure. It offers an efficient solution for conjecturing maintenance of devices for smart cities and real-time asset tracking. It combines AI capabilities. It also provides support to the wide range of latest operating systems.
   Website: Google cloud platform.
2. **Particle:** the particle is called the next-generation IoT platform. This platform is designed by spark care. This platform is compatible with IoT products and WiFi or 6G connectivity. This is the leading middleware layer platform for cellular connectivity to the cloud. This is the only online platform to bring out machines, equipment, and assets. It can give us firewall-protected clouds.
   Website: particle.io.

3. **Cisco IoT Connect Cloud:** *Cisco* IoT products provide various solutions to the IoT platforms users. This platform increased the production efficiency of the commerce sector. This platform secures the commerce sector with edge intelligence. This platform modernized the edge infrastructure. This platform also reduces network congestion for a safer transport system.
   Website: cisco.com.

4. **Amazon AWS IoT Core Platforms:** AWS core IoT will help device connectivity to the cloud. It manages all the cloud services AWS IoT connects with the cloud and interacts with the other devices and cloud applications. It provides a lightweight protocol such as MQTT and HTTP. It processes vast amounts of data and messages. It is a secure and reliable platform for transmitting messages to AWS endpoints and other services. These platforms help users to communicate and track when not connected. Users can use the different AWS services such as AWS Lambda and Amazon quick sight.
   Website: Amazon AWS IoT core.

5. **Salesforce IoT Cloud:** this platform will help to set the targeted customers, partners, devices, and sensors with relevant actions. This platform is mainly designed for digital marketing, automation, and market analytics. It allows users to test ideas for business without any programming skills. It can easily interpret the data from any device. It can give real-time traffic.
   Website: Salesforce.com.

6. **IBM Waston IoT:** this platform will provide the facility to capture and interrogate data and devices, machines, and equipment and find out the understanding for better decisions. IBM platform will work on optimization processes and resources. It can provide a bidirectional communication facility. It can have features of AI or analytics. It is the expertise of domains. It offers better security solutions and captures real-time data access.
   Website: ibm.com/Internet-of-things IBM WastonIoT.

7. **Microsoft Azure IoT Suite:** the platform is mainly designed according to industrial requirements. This platform is primarily used by manufacturing to transportation to retail. It provides solutions for intelligent spaces, remote monitoring, and connected products. This platform is very much compatible. Skilled and unskilled people can use this platform. This platform provides two solutions to open as an IoT or Sass and open-source IoT templates.
   Website: Microsoft Azure IoT suite, azure.microsoft.com.

8. **Voracity Platforms Manage IRI Voracity:** the massive, large databases. It is a fast, the adaptable platform for data searching, navigation, and reporting device data. It is a small imprint for data manipulation in rapid aggregation on the edges, it connects and integrates sensors, logs, and many more data sources. It is used for data filtering, modification, mining, and reporting. It runs on various platforms such as Linux, UNIX, and windows, from a Raspberry Pi to a Z/Linux mainframe. Universal forwarder option for agile composition and direct indexing techniques on Splunk for cloud analytics and actions on the cloud IoT data.
   Website: iri.com.

9. **Samsung Artik Cloud:** this platform has the distinct feature of interoperability. This platform dependency is based on hardware as a service. This platform depends upon cloud Monetization with IoT devices. It can have in-built security features for devices, applications, and a user-friendly environment. It keeps the connectivity data with all the devices of the cloud. In the future, it will be used in smart homes and IIoT.

    Website: Samsung Artik.

10. **ThingWorx:** it maintains the management cycle of IoT applications. It can provide features for quickly accessing data on IoT from off-premises and off-premises from the hybrid environment. ThingWorx is compatible in various aspects such as costs, visibility control, and improved connectivity IIoTs. It can be used in IIoTs and cloud applications and can access the web server's data on-premises.

    Website: ptc.com, ThingWorx.

11. **Altair Smart Works:** Altair brilliant works till the end of a process. It provides an end-to-end platform. It provides the forum as a service. It can have an open architecture. Using this platform or intelligent networks, we connect the devices such as sensors, features detection, and gateways machines. It can provide good features and functionalities.

    Website: altairsmartworks.com.

12. **Oracle IoT:** this is a much-preferred middleware platform used more efficiently. Oracle IoT platform is more important for innovation as well as animation of devices with IoT. It is simple and easy to access. Cloud systems provide a two-way connection between the device and the cloud developed with IoT technology. This platform architecture is designed to connect various devices at the same time. This platform provides a very high-security level.

    Website: Oracle IoT.

13. **SAP:** SAP IoT platforms make business technology smarter. This IoT platform uses power-driven business tools with a large volume of data and is easily managed. SAP runs intelligent business processes autonomously at the edge and orchestrates from the cloud. This platform also performs flexible and reliable message processing. SAP has included features such as IoT devices and integration services and provides scalable data for integration.

    Website: www.cloud4c.com.

14. **Itron:** Itron platform uses the products and purpose of making energy-efficient intelligent devices. This platform provides power-safe, reliable, and resilient energy and water systems. This platform offers services to the customer at a meager cost. This platform effectively manages all the resources.

    Website: www.itron.com.

15. **Predix:** Predix is an intelligent IIoT platform used in the industries for cloud formation and digital applications. This platform is used to secure and scalable IoT. This platform is used for asset and operations performance management in IIoT. This platform provides edge-to-edge connectivity, data analytics, and data processing to many IoT initiatives.

    Website: www.ge.com.

16. **Deutsche Telecom:** Deutsche telecom is the most versatile IoT platform. This platform works on the SAAS; it is dependent upon the software as a service solution. This platform is used for the transformation of classic companies into digital service providers and enables them into new business models in the easiest way. This platform is rapidly used by the automotive, manufacturing, retail, and logistics industries.

    Website: www.telekom.com.

17. **Kaa IoT Platforms:** Kaa is an open-source middleware platform. This platform has the feature to complete at the end IoT solutions, interacting application, and intelligent digital devices. The Kaa IoT platform comprises a Kaa cluster control service, bootstrap service, operation service, and third-party components such as zookeeper, SQL database, and No SQL database. Kaa IoT platforms are also called next-generation platforms because of internode communication methods.

    Website: www.kaaiot.com.

In Table 9.1, the IoT platform is taken from 2008 to 2021. Their working procedure and websites and the development procedure. Reference model for IoT middleware – IoT is done with the makeover of beautiful scenarios and presented where gadgets are used to study user requirement and act accordingly to improve experience and quality of life. Middleware platform that integrates data from several devices and acts accordingly. Due to this reason, the IoT middleware layer is present in IoT scenarios.

The middleware platform is the most intelligent and responsible part of IoT. The requirements of IoT platforms should be represented by IoT architecture. Some problems that are faced by the middleware layer on IoT are as follows:

1. **Interoperability:** the IoT is the combination of heterogeneous devices to cooperate to realize a familiar task or transmit messages to each other. The middleware layer required a platform that should be interoperable for heterogeneous things and objects predicted in future. The global sensor network is used to fulfill the requirements of middleware IoT platforms. The main drawback is that it is not compatible with all types of sensors, which can contradict the interoperability purposes.

2. **Trust:** As we know that IoT can be used in healthcare industries. Many surgeries are performed at a distance using IoT healthcare tools and devices. In these cases, any unexpected situation can occur which causes the patient death.

3. **Scalability:** IoT is a trending technology in 2020. Sixty million devices may be connected to it with the help of the Internet. The proposed middleware platform should handle the increasing number of devices and ensure that they function appropriately from different locations.

4. **Heterogeneity Abstraction:** IoT is a combination of heterogeneous things having different processing, sensing, and communication strategies of various software. They are used in many applications for multiple inputs and outputs. These middleware platforms provide a higher abstraction level to provide low-level communication complexity.

TABLE 9.1

IoT Platforms Year-Wise Development and Website

S. No.	Platform	Invented Year	Area	Website
1.	Google Cloud IoT Platform	2008	Government Healthcare, Retail, Media, Energy, Finance	Cloud.google.com
2.	Particle	2013	OTA Software Updates, Sim and device, Data Governance	Particle.io
3.	Cisco IoT Connect Cloud	2014	Mobile Service, First virtualized packet core, Virtual Network Service	Cisco.com
4.	Amazon AWS IoT Core platform	2015	Analytics services, Device Software, Connectivity control	Aws.amazon.com
5.	Salesforce IoT	2015	CRM, Cloud Service, Field Service, Community	Salesforce.com
6.	IBM Waston IoT	2015	Market-leading AI and analytics security, Dedicated Privacy	ibm.com
7.	Microsoft Azure IoT	2015	Energy Process Manufacturing Automotive Retail	azure.microsoft.com
8.	IRI Voracity	2016	Big Data Analytics, ETZ Modernization Data Governance	Iri.com
9.	Samsung Artik Cloud	2016	Device Simulator, API Console	arktik.io
10.	Thingworx	2017	Real-time visibility, Plant Benchmarking, Digital Work	ptc.com
11.	Altair Smart Works	2017	Performance optimization, Expands product life	Altairsmartworks.com
12.	Oracle IoT	2018	Asset Monitoring, Fleet Monitoring, Production Monitoring, Connected Worker, Cloud Service	oracle.com
13.	SAP	2019	SAP has provided the best services to cloud IoT platforms over the last couple of years. They offer the most trending technology, machine learning, and big data.	www.cloud4c.com
14.	Itron	2019	It is used for company specialization in renewable energy resources. It is used for resource monitoring and optimization.	www.itron.com
15.	Predix	2020	It is an IoT platform that decided to take the steps to pass the game. It provides connectivity and analytics to the aviation sector.	www.ge.com
16.	Deatsche Telekom	2021	This platform provides better connectivity. These platforms are experts in mobile IoT. They have the advantage of 5G Connectivity and the ability to update the new standards	www.telekom.com
17.	Kaa IoT Platform	2021	Kaa IoT platform provides better speed in the case of IoT implementation. It provides a comprehensive set of IoT Features. It includes onboard IoT analytics.	www.kaaiot.com

5. **Spontaneous Events:** spontaneous events occur randomly, and there should be unpredictable and abrupt connections between objects at a given time. If the randomness of the thing is increased, then mobility factors also increase.

6. **Multiplicity:** smart objects have minimal filtering and memory storage. Multiple services have been designed for a given bright object and nodes. Every node can choose and process the situation for different results and benefits. The middleware platform should be flexible, but it can partially support multiplicity.

7. **Security and Privacy:** IoT devices have mainly two types of security: operational safety and data security. Data security process exchanging things that should maintain the security policy, data integrity, nonrepudiation, confidentiality, authorization, authentication, availability, operational security, and performance, which should never lose data whether the network should be lost.

8. **Bootstrapping:** It is a crucial procedure in IoT. It ensures that intelligent things compute the network to collaborate with currently available objects. In IoT, bootstrapping should be done in an appropriated manner. Smart objects transmitted their message to their neighboring nodes that should be present. Its adjacent nodes have the responsibility that they can independently combine the network. That technique must be added to the middleware.

9. **Modularity:** Modularity is mainly connected to security. It provides two-way communication in Internet-connected things and cloud management such as Amazon AWS IoT. Modularity is a further significant component implemented in complicated applications. Middleware in IoT platforms should be managed with the help of ubiquitous computing and heterogeneous program for study similarity to provide a smooth performance which is extremely tough.

10. **Extensibility:** Around 60 million devices are connected with the IoT platforms. It is a transformative forwarding technology that is emerging. Its progression added advanced challenges to the middleware platform. It should grant the latest devices to connect independently to IoT devices. With over 60 million devices connected to IoT platforms, this transformative technology is rapidly emerging and bringing new challenges to middleware platforms. To ensure the latest devices can connect seamlessly to IoT devices, middleware platforms must continuously evolve to meet these advanced challenges. The IoT presents both challenges and opportunities for organizations and researchers in all fields. The powerful devices that interact with IoT platforms can perform tasks with precision and accuracy, delivering reliable results.

9.6   CONNECTIVITY OF IoT PLATFORMS MIDDLEWARE LAYER USING 6G TECHNOLOGY

The IoT platform and the IoT have a multidisciplinary approach to connecting with devices. The broadband connectivity is going on with geospatial and terrestrial satellites. The continuous connectivity required a high bandwidth network. The high bandwidth network is coming with the advancement of many features, so this network is called the future network. In this era, the future network is 6G and 5G. The 6G network connectivity maintains the connectivity of the IoT platform in the proper way. IoT platform connectivity with this network makes it more efficient, relevant, and accurate. The connectivity of a 6G network with IoT platforms speeds up the process and provides the best services to the user. This chapter presents a proposed structure of 6G network connectivity with the IoT platform's middleware layer. This proposed structure helps to perform more experiments with IoT platforms, their network connectivity, bandwidth, and changes in the venue. This model allows you to accomplish the performance objective, which are little packets of network and faster channel access for low latency and multidimensional links involving multiple access channel reliability and connectivity. The 6G networks have subnetworks, hyper-specialized slicing, and rancor coverage. This network has advanced paradigms regarding new security, protection, privacy, and policy. The 6G network blindly supports long-tailored solutions and mobile network features as a platform. The 6G network is compatible with IoT platforms using future state prediction, correlation recovery, proactive exploration, and self-learning. The proposed model shows the 6G network's extensive connectivity, consisting of satellites, towers, fiber optic, telecommunication orders, and IoT platforms connectivity with the network and end users with the middleware layer in Figure 9.3. This model consists of IoT platforms used by smart cars.

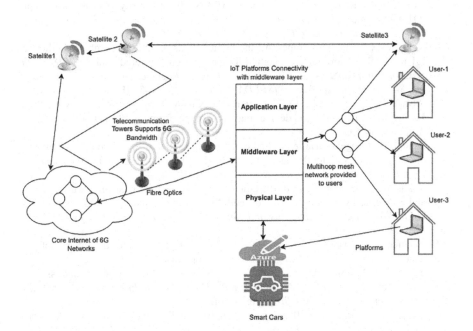

FIGURE 9.3 IoT middleware layer connectivity with 6G network Internet access procedure.

9.7 CONCLUSION

This chapter examined the main issues faced by the IoT middleware. Thus, we highlight the middleware approach depends on their functional and nonfunctional techniques and discuss the different platform approaches. This research provides an idea about the pros and cons of the existing middleware approaches. It highlights the main problem to deal with yet not well explored, such as security, privacy bootstrapping, and reliability. The middleware platform developers provide extra time to make them more user-friendly without compromising safety, as usability with a certain degree of quality might be the key to prosperity in the crowded market. In the future, the connectivity of IoT platforms is increased with the latest networks such as 7G or 8G to make the ecosystem and society smarter. In future, the industry become fully automated these fully automated industries gave the birth to Industry 4.0. In automated industries, the advanced network is required such as 6G, 7G, and 8G. These networks are installed in the industries by applying routers at different locations, RFID tags, sensors, etc. In this chapter, we have only given the idea of network connectivity, but many more experiments are performed in this regard.

REFERENCES

1. da Cruz, M.A., Rodrigues, J.J.P., Al-Muhtadi, J., Korotaev, V.V., and de Albuquerque, V.H.C., 2018. A reference model for internet of things middleware. *IEEE Internet of Things Journal*, 5(2), pp. 871–883.
2. Fersi, G., 2015 June. Middleware for the internet of things: A study. In *2015 International Conference on Distributed Computing in Sensor Systems* (pp. 230–235). IEEE.
3. Pallavi, K.N., Kumar, R., and Kulal, P., 2019. Study of security algorithms to secure IOT data in middleware. In *Proceedings of the 2nd International Conference on Green Computing and Internet of Things, ICGCIoT 2018*. VVCE.
4. Kokkonis, G., Chatzimparmpas, A., and Kontogiannis, S., 2018 September. Middleware IoT protocols performance evaluation for carrying out clustered data. In *2018 South-Eastern European Design Automation, Computer Engineering, Computer Networks and Society Media Conference (SEEDA_CECNSM)* (pp. 1–5). IEEE.
5. Silva, J.D.C., Rodrigues, J.J.P., Saleem, K., Kozlov, S.A., and Rabêlo, R.A., 2019. M4DN. IoT-A networks and devices management platform for internet of things. *IEEE Access*, 7, pp. 53305–53313.
6. Sandor, H., Genge, B., Haller, P., and Bica, A., 2019 June. A security-enhanced interoperability middleware for the internet of things. In *2019 7th International Symposium on Digital Forensics and Security (ISDFS)* (pp. 1–6). IEEE.
7. Bakhshi, Z., Balador, A., and Mustafa, J., 2018 April. Industrial IoT security threats and concerns by considering Cisco and Microsoft IoT reference models. In *2018 IEEE Wireless Communications and Networking Conference Workshops (WCNCW)* (pp. 173–178). IEEE.
8. Hoffmann, J.B., Heimes, P., and Senel, S., 2018. IoT platforms for the Internet of production. *IEEE Internet of Things Journal*, 6(3), pp. 4098–4105.
9. Yadav, E.P., Mittal, E.A., and Yadav, H., 2018 February. IoT: Challenges and issues in Indian perspective. In *2018 3rd International Conference on Internet of Things: Smart Innovation and Usages (IoT-SIU)* (pp. 1–5). IEEE.
10. Yu, J.Y. and Kim, Y.G., 2019 January. Analysis of IoT platform security: A survey. In *2019 International Conference on Platform Technology and Service (PlatCon)* (pp. 1–5). IEEE.

11. Frustaci, M., Pace, P., Aloi, G., and Fortino, G., 2017. Evaluating critical security issues of the IoT world: Present and future challenges. *IEEE Internet of Things Journal*, 5(4), pp. 2483–2495.

12. Conti, M., Dehghantanha, A., Franke, K., and Watson, S., 2018. Internet of Things Security and Forensics: Challenges and Opportunities.

13. Viswanathan, H. and Mogensen, P.E. (2020). Communications in the 6G era. *IEEE Access*, 8, pp. 57063–57074. https://doi.org/10.1109/ACCESS.2020.2981745

14. Barakat, B., Taha, A., Samson, R., Steponenaite, A., Ansari, S., Langdon, P. M., Wassell, I. J., Abbasi, Q. H., Imran, M. A., and Keates, S. (2021). *6G Opportunities Arising from the Internet of Things Use Cases: A Review Paper*, 1–29.

15. Nawaz, S.J. and Member, S. (2021). Non-coherent and backscatter communications: Enabling ultra-massive connectivity in 6G wireless networks. *IEEE Access*. https://doi. org/10.1109/ACCESS.2021.3061499

16. Jang, H. S., Jung, B. C., and Member, S. (2021). Resource-hopping based grant-free multiple access for 6G-enabled massive IoT networks. *IEEE Internet of Things Journal*, 8(20), pp. 15349–15360. https://doi.org/10.1109/JIOT.2021.3064872

17. Nkenyereye, L. and Hwang, J. (2021). Virtual IoT service slice functions for multi-access edge computing platform. *IEEE Internet of Things Journal*, 8(14), pp. 11233–11248. https://doi.org/10.1109/JIOT.2021.3051652

18. Swamy, S. N. and Kota, S. R. (2020). An empirical study on system level aspects of internet of things (IoT). *IEEE Access*, 8, pp. 188082–188134. https://doi.org/10.1109/ access.2020.3029847

10 Healthcare IoT
A Factual and Feasible
Application of Industrial IoT

Subasish Mohapatra, Amlan Sahoo,
and Subhadarshini Mohanty
Odisha University of Technology and Research

Munesh Singh
PDPM IIITDM

CONTENTS

DOI: 10.1201/9781003321149-10

10.1 INTRODUCTION

Modern social and economic trends, such as the Sustainable Agenda 2030 granted by the United States [1], aim to boost the nature of well-being conditions to increase life expectancy. Real-time implementation of this agenda has been made possible by recent technological breakthroughs. To do this, several focused technological efforts have been implemented. For instance, one such program is "Healthcare 4.0," which is being created in response to the advent of maintaining the digital well-being standard for healthcare services [2]. Additionally, recent technological advancements have made it possible to remotely and automatically monitor healthcare services using medical equipment that are designed to keep track of a patient's numerous health issues. It is important to note that these gadgets operate independently and are devoted to particular health issues. Industry 4.0, which over the previous decades had made the healthcare industry increasingly digital, gave rise to "Healthcare 4.0." For instance, X-rays, MRI, and ultrasound scans have all evolved into computer tomography and electronic medical records, respectively [3]. Caretakers and medical professionals utilize these user-centered gadgets to monitor and treat patients' medical issues as well as provide preventative care and wellness solutions. The usage of things in the Internet of Things (IoT) technology and their respective applications are expanding rapidly. As a result of recent technological advancements, such as industrial Internet of Things (IIoT) and industrial cyber-physical systems (ICPSs) [4] predicted that by the end of 2021, there will be 212 billion IoT devices and apps, with healthcare accounting for around 41% of their total usage. In addition, Grand View Research [5] estimates that by 2025, the healthcare IoT industry would be valued at USD 534.3 billion. The major motive behind this technology-driven application in healthcare is to provide better and reliable assistance along with dynamic healthcare operations by remotely and continuously regulating the patients' varied health problems [6–8].

This study has provided the essential components of HIoT (Healthcare IoT) to enable medical personalization of the traditional healthcare system as opposed to the current ones that merely support any technical assistance. Thus it offers various contributions; to promote clinical personalization, the authors contextualized HIoT (Healthcare 4.0) to define unique complete customized healthcare services (CPHS). Various measuring parameters of the HIoT are determined. As an illustration of contemporary HIoT, a use-case scenario has been provided. To establish the identified HIoT requirements in every layer of the suggested architecture, proper investigation and analysis points have been noted. Finally, dependable, robust, real-time, complete, individualized healthcare services, which solve the stated shortcomings of the current methodologies, are discussed in the framework section.

The remaining sections are structured as follows: various applications of the IIoT are discussed in Section 10.2. Section 10.3 derives some of the related works. The introduction to the personalization of healthcare IoT depicted is in Section 10.4. The analytics for the IoT-based healthcare systems, their many levels, and current initiatives to define criteria for contemporary customized services are all described in Section 10.5. Major advantages and disadvantages are derived in Section 10.6. Similarly, Section 10.7 elaborates on the suggested framework. Finally, Section 10.8 brings our chapter to a close with the conclusion and some of the future aspects.

10.2   APPLICATIONS

It is the height of automation. Every tangible item that exists becomes empowered and linked in an enterprise setting. A new generation of appealing automation is coming. The IT landscape with cloud technology has taken an important place thanks to its unstoppable ability to rapidly and simply accomplish the long-standing aim of infrastructure optimization. The integration of diverse systems and sensors is the main challenge. In order to make decisions, it is also critical to draw conclusions from streaming data from all sources. IoT applications are not constrained to a single category since they may be implemented in any type of equipment depending on the situation [9]. Electrical, mechanical, physical, and electronic components with smart labels, barcodes, LED lights, beacons, and pads are the main elements present in any IoT ecosystem. The sections that follow explain the most well-known and significant business, consumer, and individual use cases that have emerged from all the many IoT-related innovations (Figure 10.1).

10.2.1   SMART GRIDS

It is a networked gadget that measures a building or home's use of energy, water, or natural gas. Due to the effectiveness of smart grids, power providers began using them instead of traditional meters to measure overall usage.

10.2.2   SMART FLIGHTS

The fusion of artificial intelligence (AI) and IoT works together to maximize advantages while minimizing expenses for the passengers. Adding AI to jet engines, the new C-series airliner featured the geared turbofan (GTF) engine from Pratt & Whitney with 5,000 sensors that could generate up to 10 GB of data per second.

10.2.3   SMART AGRICULTURE AND FARMING

Farmers may prepare for meteorological conditions like rain, drought, snow, or wind by forecasting them in advance. Microclimate conditions may be managed using Heating, ventilation, and air conditioning (HVAC) and greenhouse environment sensors. Utilizing remote monitoring and wireless sensors, smart farming reduces the resource consumption of IoT-based pest control and smart tractors.

10.2.4   SMART MANUFACTURING

Supply chain management and intelligent logistics are both used by the IIoT.

10.2.5   SMART CITIES

Smart cities, which are described as municipalities that employ data and communication technology to improve operational efficiency, exchange information, and the standard of public services, are made possible by the IoT. The IoT can be used for the several problems such as long wait times to find parking; exceeding emissions

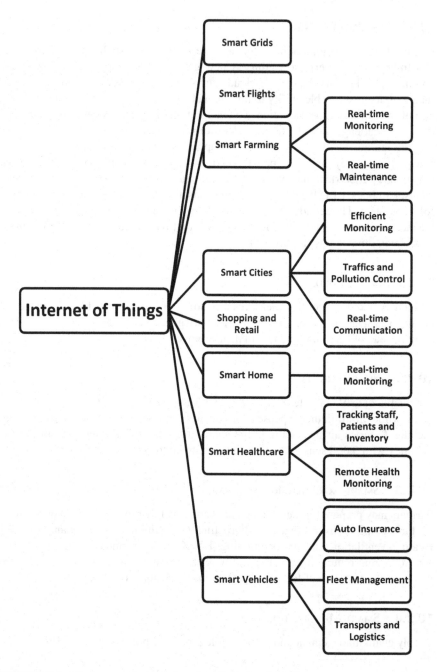

FIGURE 10.1   Various applications based on the industrial IoT.

thresholds; and unauthorized parking. To save needless emissions and traffic, one must choose to park straight away in a nearby garage rather than driving about the city center looking for an on-street spot. Better monitoring cities can automate operations that can be time-consuming or expensive, thanks to ingenious parking services and management solutions.

10.2.6 SHOPPING AND RETAIL

The IoT enables users to link their mobile devices to storefront windows, discover what they need, and receive helpful advice or perks in exchange for their loyalty.

Smart homes: If a room is empty, smart home sensors adjust the thermostat to lower the temperature and switch off the lights.

10.2.7 SMART HEALTHCARE

IoT improves supply chain management and rationalizes asset tracking. The use of an inventory management system allows individuals to locate necessary tools quickly. Patients may be able to receive remote health monitoring instead of going to the emergency department in some circumstances. The most common use in healthcare is remote health monitoring.

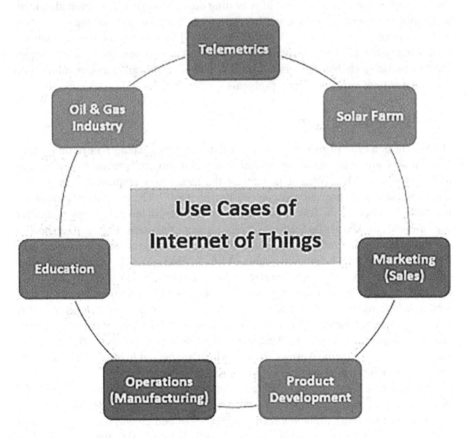

FIGURE 10.2   Various use cases of the Internet of Things.

10.2.8 Smart Vehicles

Making driving safer by using sensors, businesses created technology that incorporates sensors into smartphones to analyze driving habits. Data collection allows the app to provide training for safer driving (Figure 10.2).

10.3 RELATED WORKS

The growth of healthcare system might raise the healthcare of anybody at anytime and anywhere, therefore minimizing other limits with the rise in coverage and quality. Remote consultation, also known as teleconsultation, has gained a lot of relevance these days. This process makes use of a broadband network and bidirectional control technologies with high resolution.

There are certain gaps in the research work of the various IoT layers within the framework of the needs for inclusive customized healthcare services [10]. These resulting gaps identify research issues related to various architectural levels and accompanying essential healthcare requirements. The identified gaps result in the conclusion are consequently reliable for a patient with various diagnostic condition as the existing architecture fall short of providing dependable, resilient, individualized, and application-specific healthcare. Similarly, it fails to comprehend the connection and impact of a person's many health issues [11]. The various comparison demonstrates unequivocally that only a small number of the aforementioned techniques fully or partially enable the personalization icon of healthcare services taking into account a patient's various health problems.

10.4 HEALTHCARE IoT

Healthcare is a type of medical treatment that involves keeping an eye on patients to see if they are acting normally or abnormally and adopting preventative measures to keep them healthy. Two components of the technology employed in healthcare are, healthcare software and computers allowing it 24/7. The majority of healthcare technologies improve the way healthcare is now delivered, but ubiquitous healthcare uses a paradigm that shifts the focus from doctors to patients. Figure 10.3 shows a sample of healthcare architecture. Medical technologies are constantly crucial since it is vital in managing health and quality of life. IoT has access to health monitoring, which enhances quality while lowering costs. Patients and doctors can accurately identify the ailment with the help of smart devices [12]. The methods used now, which include mobile devices and external sensors, are archaic. When compared to other systems, healthcare systems may occasionally be the least efficient, yet there are several approaches to make them more effective. Currently, diagnostic mistakes account for 20% of medical errors, 20% of which are caused by inaccurate diagnosis methods, and 20% of which are caused by treatment delays.

By gathering analysis using patient data and medical resources, connected healthcare devices can improve patient monitoring and medical decision-making. There are connectivity problems with many types of equipment, but there are some cutting-edge items on the market that are already demonstrating the trend of IoT in

Equipped & Instrumented (Measure and sense the current condition permanently)	• Automatic information gathering, proactive condition management, and provision of preventative treatment are all features of smart health systems • A wearable device has sensors that detect physical changes like pressure, motion, and temperature
Related & Interconnected (Devices in system communicate, share data with each other)	• To allow better judgments and complete, coordinated healthcare, smart health care systems reduce informational barriers and seamlessly incorporate data and analytics into healthcare procedures • Real-time vital sign and activity monitoring equipment for use at home and on the go that connects to personal health record systems, computers, cellphones, and healthcare providers
Smart, Quick & Intelligent (able to predict and to respond on unplaned events)	• Intelligent health systems continuously analyse data from several devices and other sources to develop insights and suggestions for the personal health regimens of the user • Analytics programmes track device data, compare it to targets, monitor goal progress, and deliver alarms as necessary using rules and logic

FIGURE 10.3   Characteristics of the smart healthcare solution.

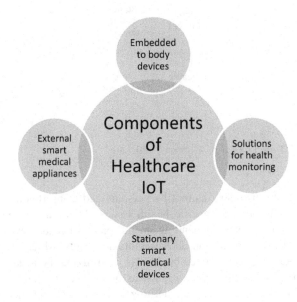

FIGURE 10.4   Components of healthcare IoT.

healthcare. The IoT components in healthcare are shown in Figure 10.4. According to current customized healthcare is a service that adapts medical care to specific people by identifying common traits as such genetics, heredity, and lifestyle [13]. The patient's total health, which necessitates an awareness of the biological relationships

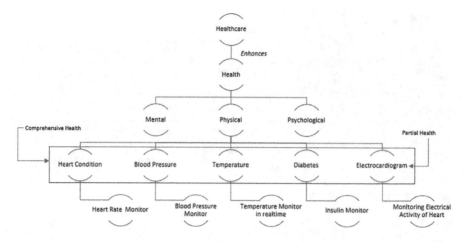

FIGURE 10.5 Illustration of partial and comprehensive healthcare architecture.

FIGURE 10.6 IoT used scenarios in smart health care.

among many health disorders, cannot be personalized using this approach, which only assistances to personal specific health issues. Additionally, the specifics of personalized healthcare vary depending on the patient because every person has a different set of health condition determinants and traits, which are essential for maintaining a person's personalized health. By minimizing the negative impacts of the patient's many medical problems (such as heart disease, high blood pressure, or insulin resistance), element-based optimization seeks to enhance the patient's overall long-term health. Establishing a comprehensive customized healthcare service for patients and enhancing their enduring health may be made feasible through Healthcare 4.0 (Figures 10.5 and 10.6).

10.5 IoT ANALYTICS IN HEALTHCARE

We live in a world where data are more valuable than the rarest minerals and more lethal than the most powerful weaponry. But it is essentially useless in its unprocessed state. It must first be treated before it can be used. There are sensor, network, and decision layers in a typical IoT design. Data from the environment are collected by the sensor layer and sent to the processing layer across the network [14]. Data analysis is carried out in this instance to assist the decision layer in drawing judgments. Data analytics is hence the IoT's foundation. IoT researchers have always found the healthcare industry to be an attractive business. Additionally, the global lack of medical experts has fueled the expansion of such studies.

The following reasons for the numerous healthcare procedures that may be optimized as a result of a strong and rapid data analysis of an IoT system:

Disease diagnosis: the physiological data, including blood pressure, body temperature, and others, acquired by wearable IoT devices can be utilized to diagnose diseases faster, in addition to other test results (urine tests, blood tests, magnetic resonance imaging [MRI] scans, etc.).

Health prevention and lifestyle management: the treatment of a chronic patient's lifestyle is aided by routine analysis of data about behavioral elements of a person, such as sleep patterns, eating habits, and daily physical activity. A healthy person may be informed of the diseases they may be at risk for owing to their present lifestyle by such an examination.

Medical emergency: if an abnormality in the physiological data is quickly identified, an emergency (such as an old person falling and fainting in his or her home) can be effectively managed. The elderly and those with terminal illnesses will benefit the most from this (Figure 10.7).

Remote real-time health monitoring: real-time health monitoring is one of the main analysis applications in the healthcare industry. IoT sensors gather and transmit the patients' physiological data. The data are then immediately examined. Every time a deviation from the usual pattern is found, a thorough report is sent to the patient's healthcare provider for additional investigation.

To enable IoT systems and applications to behave adaptively, these collected and aggregated insights may be looped back into them. In other words, data-driven insights enable hardware and software to act intelligently. IoT data must be meticulously collected since it conceals a range of useful patterns, correlations, recommendations, information, knowledge, etc.

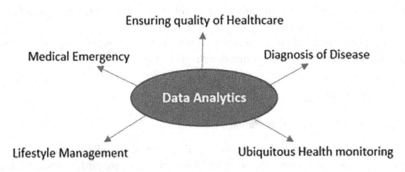

FIGURE 10.7   Utility of IoT-based data analytics in healthcare.

10.6   BENEFITS

10.6.1   REAL-TIME REPORTING AND MONITORING

In the case of a medical emergency, real-time monitoring via associated devices can save a million lives. Connected devices can gather important medical and health-related data when real-time condition monitoring is implemented using a smart medical device connected to a smartphone app. The IoT-connected gadget gathers and transmits health data, including measurements of various body parameters. The information is kept in the cloud and can be sent to a person who has been permitted to view it which is made available ubiquitously.

10.6.2   END-TO-END CONNECTIVITY AND AFFORDABILITY

Through healthcare mobility solutions, IoT may assist in transferring the healthcare data, which greatly reduces the cost of a result, and this technology-driven arrangement can reduce expenses by reducing needless trips and employing resources of higher quality enhancing resource allocation and planning throughout the process.

10.6.3   DATA ASSORTMENT AND ANALYSIS

It is not as simple as it seems for healthcare practitioners to manage a large volume of data. Through mobility solutions driven by IoT, real-time data gathered by IoT-enabled mobile devices may be analyzed and segmented.

10.6.4   TRACKING AND ALERTS

With continual notifications and real-time alerts for appropriate analysis, tracking and alerts in life-threatening circumstances might prove to be a lifesaver to protect a crucial patient's health. Real-time tracking, notification, and monitoring are made possible by IoT-powered healthcare mobility solutions.

10.6.5   REMOTE MEDICAL ASSISTANCE

When a patient needs medical care but is unable to reach a doctor because of obstacles like distance or ignorance, it is a horrible scenario.

Through healthcare delivery networks connected to patients through IoT devices, patients can take prescription medications right at home [15].

10.7   CHALLENGES

10.7.1   DATA SECURITY AND PRIVACY

This is a major issue that IoT is now facing. Real-time data collection is not possible unless it is complying with data standards with protocols. Regarding ownership and control of data, there is a lot of confusion. False health claims and the fabrication of phony IDs for prescription purchases and sales are two instances of how IoT device data have been misused.

10.7.2 INTEGRATION OF MULTIPLE DEVICES AND PROTOCOLS

The application of IoT in the healthcare industry is hampered by the integration of many device types. This obstacle exists because device makers cannot agree on a communication protocol with a set of standards.

10.7.3 DATA OVERLOAD AND ACCURACY

The inconsistency of various methods makes it challenging to aggregate data for critical understanding and analysis. IoT gathers data in large quantities, thus for efficient data analysis, the data need to be divided into manageable chunks without being overloaded with fine precision.

10.7.4 COST

IoT app development for healthcare mobility solutions, costs are one of the bigger difficulties. However, if the IoT solution addresses a real problem, the expenses are justified. Even though the investment of a lot of resources and money into developing an IoT application, the benefits will be enormous when a company uses IoT to improve business processes, generate more income streams, and open up new business options while also saving time and labor costs.

10.8 SUGGESTED FRAMEWORK

The terms "Internet" and "things" are combined to form the phrase "Internet of Things" (IoT). Smart gadgets, instruments, and ubiquitous, context-aware things that can communicate are all referred to as things since they can connect to the internet. For IoT to be truly widespread, IoT devices need to have the following three fundamental characteristics:

- **Sensing Capability:** IoT devices need to be able to perceive and integrate the data they collect. For instance, numerous specialists are used in the healthcare industry to sense and collect data on various body parameters. The data collection is self-contained.
- **Communicable:** IoT devices should be able to transmit data to the various desired data centers after sensing and aggregating using a variety of communication mediums, such as wireless technologies, mobile networks, Wireless Local Area Networks (WLANs), Wireless Sensor Networks (WSNs), and Mobile Ad Hoc Network (MANET), among others.
- **Actionable:** The aggregated data are meaningless on their own, thus for IoT devices to respond, they must be able to comprehend the data. Until processing technology is utilized to detect whether a body parameter level exceeds or deviates from the normal range.

In this process, three-tiered HIoT architecture has been taken into consideration. The things layer comes first in this three-tier system, trailed by the communication and application layers.

- **Things Layer:** This stage in the HIoT refers to sensing systems and apparatus that, depending on the application, record the different values seen from various sources (for example, apparatus monitoring patients' various body parts or health issues). These sensors are designed to gather data from a patient's body related to their ailment. The devices and sensors mentioned above collect information from the patient's body and, in some cases, send it to the data center. Some responsive sensors have devices that keep track of the relevant body parameters.
- **Communication Layer:** The communication layer is viewed as the foundation of IoT systems. IoT devices generate a substantial amount of continuous data that are challenging to transmit and store. All of the data are sent from the things layer to the application layer through the communication layer. An appropriate communication medium is required to convey data from all of the Things layer nodes to the other layers of the system. The communication medium may be wireless (WiFi, Bluetooth, RFID, 5G, etc.) or wired, depending on the communication protocol developed (Ethernet, USB, etc.).
- **Application Layer:** The application layer focuses on delivering top-notch healthcare services and directly connects the end user to the cloud-based IoT platform. HIoT is a network of interconnected hardware, software, devices, and services. Their use is expanding tremendously as a result of network improvements. These are employed by healthcare workers to process and analyze data to make decisions and treat patients. The IoMT in hospitals often focuses on enhancing patient safety and/or streamlining procedures. It makes it possible for practitioners to collaborate across disciplinary lines to provide tailored patient care. The application layer serves two purposes by offering: On the one hand, monitoring the body parameters and aiding in the supervisory decisive outcome; on the other, support applications serve as user interfaces for a variety of health monitoring equipment. To enable tailored healthcare services, many methodologies have been created.

To achieve the objective of the overall system there is a fusion made between wearable and mobile technology. When needed, the system provides both residents with cognitive difficulties with immediate help through their wearable gadgets. Automatic alerts are sent to carers in potentially dangerous circumstances. The discussed methodology consists of different elements to guarantee the operational dependability of correlated healthcare IoT services and applications to provide individual services. Figure 10.8 depicts the workflow of the personalized HIoT system for all-encompassing personalized healthcare. Data are gathered at the IoT cloud and delivered to the concerned predictive model, where it is stored, from various monitoring devices used by patients, such as heart rate, temperature, blood pressure monitor, etc. Data

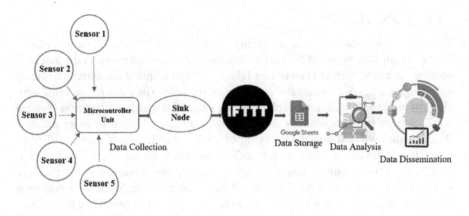

FIGURE 10.8 Flowchart of the suggested framework.

that have been stored are processed using various predictive methods and disease models. To create the personalized HIoT system for all-encompassing personalized healthcare, there are several obstacles to overcome.

Additionally, the monitor offers clinical reasoning to precisely pinpoint the origin of the disparity. Once the disparity has been located, the monitor may treat it appropriately, and the treatment's results can then be observed. The monitor remotely and automatically manages an individual's numerous physical conditions based on clinical features. The clinical medication enables the monitoring process to be self-conscious, which promotes the provision of healthcare services free from incorrect diagnoses.

Modeling health problems is a difficult undertaking since it calls for determining the many clinical relationships between the various patient health states. However, these dependencies are frequently complex and rely on knowledge of the circumstances' underlying variable-level biological processes, which are mostly unknowable. To focus the various healthcare services based on their clinical features, a library of distinct health problems (models) will be created that may subsequently be employed in any computerized healthcare system. The library will also aid in discovering further connections between multiple fitness disorders which are normally outside the scope of concerned professionals. Additionally, based on the models, it will be possible to describe the newly discovered dependencies in a way that both practitioners and machines can comprehend. Similarly, predictive approaches can be used to compare the data of many patients with comparable illnesses to better comprehend health issues, leading to the recommendation of a more individualized prognosis and therapy that will efficiently enhance patients' health. In several facets of healthcare, AI is essential. AI is now employed for data gathering, communication, and monitoring at the things, communication, and navigation levels, respectively. Due to its expandability and trust concerns, machine-based prediction is not yet totally safe for diagnostic and treatment purposes. As a result, essential healthcare fields do not yet completely trust smart diagnosis systems.

10.9　CONCLUSION

Numerous industrial segments are seeing advancements because of the Internet of Things. Healthcare is one of the sectors that has embraced this prospect the quickest, opening up a new market-centered on HIoT. This fact inspired the authors to create a thorough survey to evaluate the condition of the field. The most recent HIoT articles and products were found, explained, and examined to achieve this aim. Several HIoT services and apps address societal demands but are becoming more isolated. Based on the collected sensor data, healthcare IoT is a collection of various IoT technologies and devices used to support individual health. Healthcare IoT offers individualized services using a single person's gadgets. Users require a comprehensive system to fulfill the strict requirements of health monitoring. Since the integration of diagnostic and therapeutic processes is crucial, information should be maintained in a manner that is simple to access and convey. An effective IoT framework for the healthcare environment describes how to combine patient information for caregivers and report patient progress using the technologies now employed in the healthcare system. Patients' physical and mental health may improve as a result of less expense and in less time.

Additionally, healthcare professionals, experts, and the general public with an interest in HIoT can use this document as a resource for information. However, this analysis does not offer a thorough grasp of several essential subjects, such as HIoT topologies, architectures, and platforms, as well as the requirements, difficulties, and suggested security models. Other technologies, such as big data, augmented reality, and cognitive systems, which are not included in this analysis but might be investigated further, exist.

REFERENCES

1. Lee, B. X., Kjaerulf, F., Turner, S., Cohen, L., Donnelly, P. D., Muggah, R., & Gilligan, J. (2016). Transforming our world: implementing the 2030 agenda through sustainable development goal indicators. *Journal of Public Health Policy*, 37(1), 13–31.
2. Gupta, R., Bhattacharya, P., Tanwar, S., Kumar, N., & Zeadally, S. (2021). GaRuDa: a blockchain-based delivery scheme using drones for healthcare 5.0 applications. *IEEE Internet of Things Magazine*, 4(4), 60–66.
3. Bercovich, E., & Javitt, M. C. (2018). Medical imaging: from roentgen to the digital revolution, and beyond. *Rambam Maimonides Medical Journal*, 9(4), e0034.
4. Giri, A., Dutta, S., Neogy, S., Dahal, K., & Pervez, Z. (2017, October). Internet of Things (IoT) a survey on architecture, enabling technologies, applications and challenges. In *Proceedings of the 1st International Conference on Internet of Things and Machine Learning* (pp. 1–12).
5. Hanson, K., & Goodman, C. (2017). Testing times: trends in availability, price, and market share of malaria diagnostics in the public and private healthcare sector across eight sub-Saharan African countries from 2009 to 2015. *Malaria Journal*, 16(1), 1–16.
6. Ramson, S. J., Vishnu, S., & Shanmugam, M. (2020, March). Applications of Internet of Things (IoT)–an overview. In *2020 5th International Conference on Devices, Circuits and Systems (ICDCS)* (pp. 92–95). IEEE.
7. Zhan, K. (2021). Sports and health big data system based on 5G network and Internet of Things system. *Microprocessors and Microsystems*, 80, 103363.

8. AlShorman, O., AlShorman, B., Alkhassaweneh, M., & Alkahtani, F. (2020). A review of internet of medical things (IoMT)-based remote health monitoring through wearable sensors: a case study for diabetic patients. *Indonesian Journal of Electrical Engineering and Computer Science*, 20(1), 414–422.

9. Vyas, S., Gupta, M., & Yadav, R. (2019, February). Converging blockchain and machine learning for healthcare. In *2019 Amity International Conference on Artificial Intelligence (AICAI)* (pp. 709–711). IEEE.

10. Vergutz, A., Noubir, G., & Nogueira, M. (2020). Reliability for smart healthcare: a network slicing perspective. *IEEE Network*, 34(4), 91–97.

11. Zhang, J., Li, L., Lin, G., Fang, D., Tai, Y., & Huang, J. (2020). Cyber resilience in healthcare digital twin on lung cancer. *IEEE Access*, 8, 201900–201913.

12. Yan, C., Duan, G., Zhang, Y., Wu, F. X., Pan, Y., & Wang, J. (2020). Predicting drug-drug interactions based on integrated similarity and semi-supervised learning. *IEEE/ACM Transactions on Computational Biology and Bioinformatics*.

13. Singh, D., Mishra, P. M., Lamba, A., & Swagatika, S. (2020). Security issues in different layers of IoT and their possible mitigation. *International Journal of Scientific & Technology Research*, 9(4), 2762–2771.

14. Qadri, Y. A., Nauman, A., Zikria, Y. B., Vasilakos, A. V., & Kim, S. W. (2020). The future of healthcare internet of things: a survey of emerging technologies. *IEEE Communications Surveys & Tutorials*, 22(2), 1121–1167.

15. Fani, M., Rezayi, M., Meshkat, Z., Rezaee, S. A., Makvandi, M., & Angali, K. A. (2020). A novel electrochemical DNA biosensor based on a gold nanoparticles-reduced graphene oxide-polypyrrole nanocomposite to detect human T-lymphotropic virus-1. *IEEE Sensors Journal*, 20(18), 10625–10632.

11 IoT-Based Spacecraft Anti-Collision HUD Design Formulation

Bibhorr
IUBH International University

CONTENTS

11.1 INTRODUCTION

The Earth's orbit is becoming increasingly congested. Reusable rockets and increasingly powerful smallsats are decreasing the obstacles to accessing and using space. In the previous decade, the cost of commercial launches has decreased by a factor of 20 [1]. Increased space activity has resulted in a significant increase in the number of objects in space and this is only expected to grow exponentially in the future. As the number of objects in Earth's orbit grows, the chances of a satellite colliding with space junk or another satellite are increasing [2]. The distribution of space objects whether vehicles, satellites, debris, etc. is also varied with some areas in the space with a high density of these space objects while others with moderate to low density. The density of space traffic elements is therefore a crucial factor in determining the traffic scenario and subsequent maneuvering of vehicles. In practical applications of AI-controlled spacecrafts and trajectory planning integrated autonomous systems, maneuvering these in an area with obstructions in the path is challenging and requires advanced capability in these systems other than the very basic feature of ability to avoid collisions with other space elements, viz., other spacecrafts, debris, or space devices orbiting for any other experimental purposes. Increased space activity can significantly aggravate the space debris situation [3]. After 60 years of space activities, the orbital population appears to be unevenly distributed, with more than 90% of the recorded objects being nonfunctional i.e., orbital garbage. Furthermore, a large number of active spacecrafts lack a propulsion engine and hence are unable to maneuver to prevent accidents [4].

Due to technological and political concerns, the space traffic control model has historically been presented in a vague, confusing, and imprecise way [5]. The notion of space traffic management is complicated. It is critical to design a model capable of successfully and efficiently regulating flightpath for collision prevention before constructing a comprehensive space traffic management system [6].

With an increased number of space objects, the collision probability has increased manifold. Because of the growing quantity of space objects, space is becoming increasingly congested. It is a given fact that the number of objects in space surrounding the Earth will continue to grow in the near future, increasing the likelihood of a collision, communication failure, or other undesired occurrences [7]. To mitigate the collision risk, it is therefore crucial to assess the collision probability and associated collision risk with high precision and much before the elements come close enough to appropriately maneuver and change the trajectories of the involved controllable elements.

11.2   RELATED WORK

Min et al. [8] detailed a collision avoidance method for satellites with formation flight highlighting the formation initialization and formation reconfiguration operations. Omar et al. [9] discussed a spacecraft collision avoidance technique employing aerodynamic drag. In order to avoid collision with inactive space objects, Lee et al. [10] proposed a semianalytic and suboptimal guidance control for an active spacecraft. Wang et al. [11] discussed a two-spacecraft collision avoidance problem. Hobbs et al. [12] proposed four types of collision avoidance. Similar other lone methods and techniques are being researched and analyzed rarely highlighting artificial intelligence and machine learning. Such works are all lone techniques based on conditionalities and conventionally propagated models, lacking the desideratum elements of futuristic aspects and parameters involving artificial intelligence, machine learning, and Internet of things. Moreover, such works are not able to typically establish inclusive solutions for the collision avoidance and management to incorporate multifaceted models/scenarios. Hence, this work has been incepted to put forth multimodel solutions and all-inclusive approach to manage space traffic accidents in an effective way irrespective of engineering conditionalities.

11.3   CYBER-PHYSICAL NETWORKS IN SPACE

Aerospace industries are aspiring to station data center networks into the Earth's orbit and in deep space, the reason for which is progressing feasibility as the cost of manufacturing and launching satellites is decreasing; they are eyeing on new business models by accomplishing this concept. The space industry aspires to replicate the IT sector's approach of colocation services regarding server hardware, edge

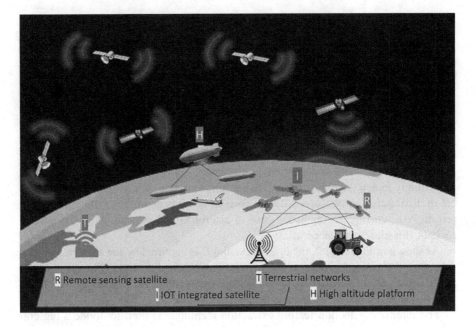

R Remote sensing satellite   T Terrestrial networks
        I IOT integrated satellite   H High altitude platform

FIGURE 11.1   Cyber-physical networks in space.

computing, running applications remotely in the cloud, etc., founded upon satellite-based futuristic models.

The work takes into consideration this model where a heavy server-mounted satellite is theoretically hypothesized to interact with other satellites to gather and analyze their data and conduct overhead edge computing. This model guarantees the Earth Observation (EO) satellite data to be distributed, stored, and processed into instantly useable images and then sent directly to grass root users. Figure 11.1 depicts such a cyber-physical network that comprises the integration of IoT-integrated satellites, AI-enabled high-altitude platforms, terrestrial network communication, ground communication, remote-sensing satellite networks, communication, and astronomical satellite networks.

11.4   SPACECRAFT ANTI-COLLISION SYSTEM DESIGN

A collision avoidance system is a safety technology that is meant to avoid or mitigate the severity of an accident during the occurrence of the accidental circumstances. The spacecraft anti-collision system (SACS) is based on the concept of establishing data centers in space, as shown in Figure 11.1. Anti-collision system design constitutes five design subsystems, viz., traffic identification design (TID), density evaluation model (DEM), trajectory planning model (TPM), risk assessment model (RAM), and finally heads-up display (HUD) system.

11.4.1 Traffic Identification Design

The traffic identification design uses a video imaging process based on an IoT-integrated video camera docked with a mathematical computing system. The video camera sends raw image data feed as a first source impression for the system to further handle the computational processes. The system works by recording image frames and sending them to the computer unit. The video feed gets processed via a series of computers and the resultant information of the visual silhouettes detected in the form of traffic elements is transmitted over the controller area network (CAN) bus that also gets displayed on the HUD system. The system's performance is predicted to be high, and it is dependent on the image analysis algorithm, which should accordingly match the onboard hardware implementation. The system must be capable of distinguishing between different sorts of barriers as well as different types of terrain, viz., corrugated (as in the case of asteroids) and smooth (as in the case of rockets, missiles, and spacecrafts).

As shown in Figure 11.2, the original visual B is captured by the camera and then subsequently scanned resulting in output A. The scanned data are communicated to the IoT-integrated satellite networks, together establishing data center C, which can later be retrieved in case of any incident sparks.

The intensity reduction feature extraction technique is used to divide a big collection of raw data into smaller, more tractable data sets for subsequent processing. These huge data sets include a great number of variables, requiring a significant

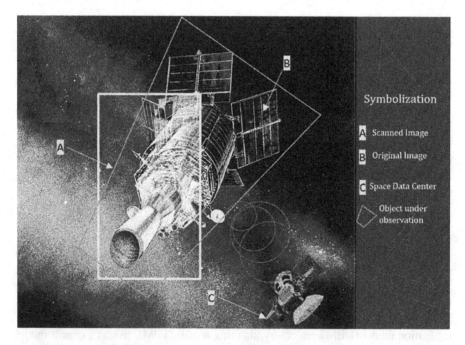

FIGURE 11.2 An illustration depicting traffic identification design where the image is first scanned, and the visual elements are recognized postcomputing.

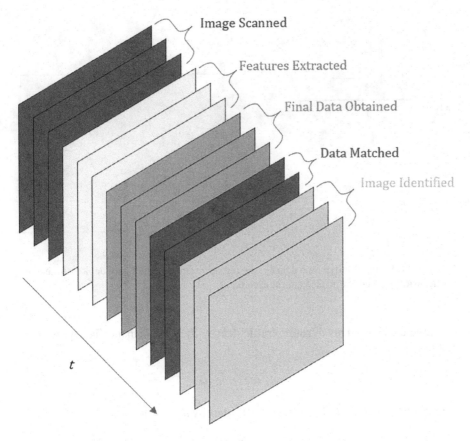

FIGURE 11.3   Progression at various layers in computing processes over a time period (t).

amount of processing resources to compute. When there is a huge data collection and it is needed to decrease the amount of physiognomy without accommodating any loss of critical or relevant information, extracting the features is a handy strategy. Most of the information in the original set of features is summarized through a new and smaller collection of features, which are further classified.

Figure 11.3 shows the progression at various layers in the computing process, which initiates from the point of image scanning and results in the identification of the image. Meanwhile, the features are extracted and the data obtained are matched with predefined elements and parameters.

11.4.2   DENSITY EVALUATION MODEL

A three-dimensional dilated spatial frame consisting of n traffic elements is rendered. The frame that encapsulates the elements has magnitude a as three of its dimensions. As the density is dynamic in nature, it changes over time t. The traffic density is a function of time, magnitude of three-dimensional spatial cube, and number

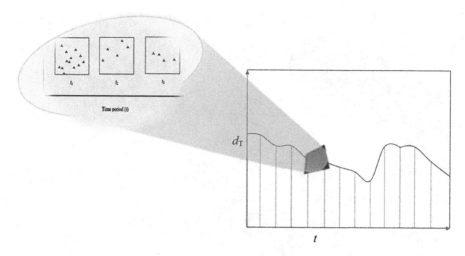

FIGURE 11.4 Plot for dynamic density; the region zoomed in the figure depicts three-time frames with reduced traffic elements at each point.

of traffic elements. This dynamic traffic density d_T is mathematically expressed as follows:

$$d_T = f(t, a, n) \tag{11.1}$$

The formula for evaluating dynamic traffic density d_T is as follows:

$$d_T = \frac{n + \Delta n}{a^3} \tag{11.2}$$

where Δn is the change in traffic elements observed over a period of time t.

The dynamic density d_T is visualized, to the pilot, in HUD screen in the form of a continuously varying graphical plot whose Y-axis denotes the density variations and X-axis denotes the time. This graphical plot is depicted in Figure 11.4.

11.4.3 RISK ASSESSMENT MODEL

The loss of human life or injuries, environmental damage, and economic losses are the primary hazards associated with spacecraft accidents. The consideration of risk studies is valuable in the preparation of strategic tools for developing and deploying emergency control and management scenarios in space.

In this model, the maximum collision probability risk $P(C)$ is given as follows:

$$P(C) = \frac{n_a}{n_t} \tag{11.3}$$

where n_a implies the number of traffic elements tested for collision and n_t indicates the total number of traffic elements present in the rendered frame.

All n_a elements account for maximum collision risk since they confirm their presence in the congested zone. The total array of elements present in the congested zone is given as follows:

$$L = \{e_1, e_2, e_3, e_4, e_5, \ldots, e_n\} \tag{11.4}$$

while the total of all the elements present in the frame is indicated as follows:

$$M = \{e_1, e_2, e_3, e_4, e_5, e_{5+k}, e_{5+2k}, \ldots, e_{5+mk}\} \tag{11.5}$$

Subtracting n_a from n_t results in n_e depicts the number of traffic elements that remained absent from being present in the congested zone.

Further to this, combination-based risk assessment probability can also be computed for precise risk possibilities. The combinations could be generated using the following formula:

$$_nC_r = \frac{n!}{r!\,(n-r)!} \tag{11.6}$$

where $n = n_a$ and $r < n_a$

11.4.4 SAFE TRAJECTORY PLANNING MODEL

As shown in Figure 11.5, d_t and D_t are the diagonals emanating from N_0 and touching the vertices of the overview projection representing the congested zone.

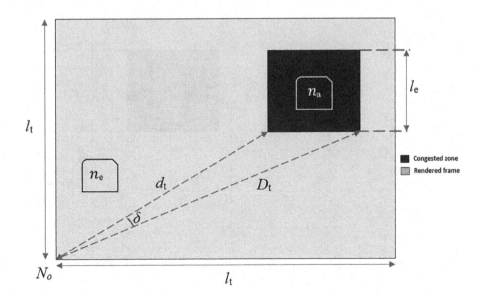

FIGURE 11.5 Depiction of traffic elements n_a and n_e in the rendered frame.

The empty zone is located by identifying the free-from-traffic (FFT) vertices at d_t and D_t. The empty zone z_e guarantees the zero-traffic movement for time period T_0 sufficient enough to maneuver the spacecraft in the safe flight zone z_s.

The distension array of empty zone is represented as follows:

$$Z = \{x, y, z\} \tag{11.7}$$

As shown in Figure 11.6, four FFT vertices are identified, two along d_t and two along D_t. An imaginary quadrilateral simulation is then established by extending lines through vertices and aligning the rest dimensions. Post the establishment of the empty zone z_e, safe zone z_s is computed using the following formula:

$$z_s = (z_e) \tag{11.8}$$

Z_s and Z_e denote safe and empty zone, respectively.

$$\text{where } f(z_e) = \sqrt[k]{ze} \tag{11.9}$$

Here, k denotes the diminishing factor.

Finally, the distension array of safe zone is obtained as follows:

$$S = \{x_n, y_n, z_n\} \tag{11.10}$$

The safe trajectory t_s is confined in the safe flight zone z_s. The safe trajectory is a function of time period T_{0+J} where T_{0+J} signifies the time period that guarantees that

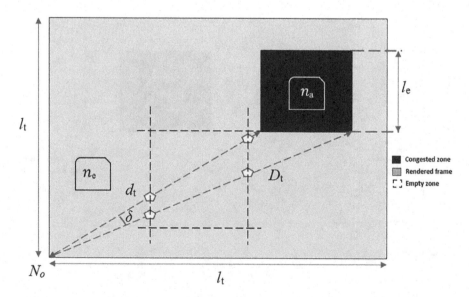

FIGURE 11.6   Estimation of empty zone in the rendered frame.

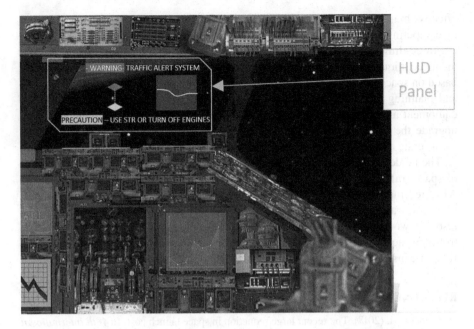

FIGURE 11.7   Heads-up display design within the spacecraft's windscreen.

the spacecraft would maneuver uninterruptedly through the safe flight zone path until the clearance of all the associated risks.

11.4.5   HEADS-UP DISPLAY SYSTEM

HUD system highlights the transparent display feature mechanism that is visually layered within the windscreen of the spacecraft and its function is to display critical information and data directly in the vision feed of pilot without distracting the pilot from the usual vision angle onto the screen. As shown in Figure 11.7, HUD is used to display a warning message in the red highlighted screen zone and the computer suggestions as a precautionary move in the green highlighted zone. The HUD is also shown displaying the dynamic density plot and the distance between the closest approaching element and the spacecraft being piloted.

In this model, the HUD system receives direct feed from RAM, post which a precaution message pops up providing recommendation strategies for the pilot to choose. The ultimate decision rests with the pilot. This has been formulated deliberately to reduce complete machine interventions so that the pilot gets complete freedom to operate.

11.5   CONCLUSION AND FUTURE SCOPE

An IoT-based spacecraft anti-collision HUD design formulation is incepted for its successful implementation in a real-space environment to help mitigate threats arising out of space traffic accidents. The design constituted five subsystems for the effective management of space traffic. The model focuses on safe trajectory planning by computing a safe zone from the previously constructed empty zone present inside the three-dimensionally distended frame. The model also presented a risk assessment technique for maximum risk evaluation and for computing risk probability based on spacecraft combinations. The model proves trustworthy for assessing risks and planning trajectories with multiple possibilities of integration with IoT-based equipment and systems without engineering conditionalities. Work is underway to upgrade the performance scenario of the anti-collision system design and to also incorporate additional mathematical models under a laboratory test environment.

The model incepted in this research could be used for modeling and simulations of space traffic management in real-world scenarios. The research delves into the AI-rested mathematical models for assessing the dangers and risks of space traffic crashes and extends the research into the field of aerospace engineering. The research would serve as a foundation for future research on preventing and controlling space traffic collisions and accidents. This research also provides new and futuristic insights into space traffic management highlighting the all-inclusive model.

REFERENCES

1. Jones, H. (2018). The recent large reduction in space launch cost. In *48th International Conference on Environmental Systems*, pp. 1–10, 8–12 July 2018, Albuquerque, New Mexico.
2. Patera, R. P. (2001). General method for calculating satellite collision probability. *Journal of Guidance, Control, and Dynamics*, 24(4), 716–722.
3. Muelhaupt, T. J., Sorge, M. E., Morin, J., & Wilson, R. S. (2019). Space traffic management in the new space era. *Journal of Space Safety Engineering*, 6(2), 80–87.
4. Bonnal, C., Francillout, L., Moury, M., Aniakou, U., Dolado Perez, J.-C., Mariez, J., & Michel, S. (2020). CNES technical considerations on space traffic management. *Acta Astronautica*, 167, 296–301.
5. Johnson, N. L. (2004). Airspace complexity and its application in air traffic management. Space traffic management concepts and practices. *Acta Astronautica*, 55(3–9), 803–809.
6. Ailor, W. H. (2006). Space traffic management: Implementations and implications. *Acta Astronautica*, 58(5), 279–286.
7. Cukurtepe, H., & Akgun, I. (2009). Towards space traffic management system. *Acta Astronautica*, 65(5–6), 870–878.
8. Min, H., Guoqiang, Z., & Junling, S. (2010). Collision avoidance control for formation flying satellites. In *AIAA Guidance, Navigation, and Control Conference*, 8–11 August 2011, Portland, Oregon.
9. Omar, S. R., & Bevilacqua, R. (2019). Spacecraft collision avoidance using aerodynamic drag. *Journal of Guidance, Control, and Dynamics*, 43, 1–7.
10. Lee, K., Park, C., & Park, S.-Y. (2014). Near-optimal guidance and control for spacecraft collision avoidance maneuvers. In *AIAA/AAS Astrodynamics Specialist Conference*, 4–7 August 2014, San Diego, CA.
11. Wang, S., & Schaub, H. (2008). Spacecraft collision avoidance using coulomb forces with separation distance and rate feedback. *Journal of Guidance, Control, and Dynamics*, 31(3), 740–750.
12. Hobbs, K. L., & Feron, E. M. (2020). A taxonomy for aerospace collision avoidance with implications for automation in space traffic management. In *AIAA Scitech 2020 Forum*. https://doi.org/10.2514/6.2020-0877

12 Coverage of LoRaWAN in Vijayawada
A Practical Approach

P. Saleem Akram and M. Lakshmana
Koneru Lakshmaiah Education Foundation

CONTENTS

12.1 INTRODUCTION

LoRaWAN is a low-power, broad-area network standard (LPWAN). Concerning the OSI levels, level 2 will be preferred as the MAC (media access control) layer. In other words, LoRaWAN is responsible for combining the channels and connection settings of many LORA devices with respect to channel, bandwidth, encryption of data, etc. (Bourennane et al., 2020). This technology enables point-to-point information to be sent and received. A LoRa device is characterized by a long range with a minimal component. It employs the spread spectrum approach, which enables many signals to be received at the same time while at the same time having varying speeds.

DOI: 10.1201/9781003321149-12

It delivers an appealing mix of properties such as ultra-long range, low-power consumption, and minimum costs, as well as secured data transfer. LoRa can give far more coverage than cellular networks, both for public and private networks.

Therefore, for its implementation in the Industrial Internet of Things (IIoT), the long-range and low-power character of LoRa is essential to enhance our standards of life swiftly. It answers many new issues, including climate change, pollution, and natural catastrophes, both in rural and urban locations worldwide (Sinha et al., 2017). LoRa is extensively used for transport, production, facilities, home equipment, and even wearable products due to its comfort and lucrative attribute, like LPWAN.

IoT generally offers several technology options for data transfer across the network, such as LoRa, NB-IoT, WiFi, Bluetooth, Zigbee, and Sub-1 GHz. However, extended transmission lengths and energy conservation in wireless network systems cannot be guaranteed at the same time because more energy is always required for a greater transmission distance. Therefore, such LoRa is the ideal solution for sending a comparably small quantity of data over a long distance so that both long range and low power are available (Butun et al., 2018).

12.1.1 HARDWARE SETUP

Here, we took the help of hardware setup of SX1272 chipset, which has been developed by Semtech company through which SX1272 module was developed. Its transmission power is 14 dBm and sensitivity −134 dBm.[1] In our practical approach, we have taken help of the SX1272 LoRa module with 4.5 dBi Antenna.[2] This module relates to Socket 0, which is placed in SX1272 Waspmote board.

This chart, which was in Table 12.1 (see note 1), illustrates the comparison between LoRa and many other commonly used wireless network technologies to

TABLE 12.1
LoRa Comparison to Other Wireless Technology

Wireless Network Technology	Velocity	Distance	Energy Consumption	Application	Communication Cost	Construction Cost
LoRa	Slowest	Longest	Low	Outdoor sensors	Free	Median
NB-IoT	Median	Short	High	Indoor sensors	Data flow charge	Low
3G/4G	Fastest	Long	High	Call and internet	Data flow charge	High
WiFi	Fast	Short	Highest	Home network	Free	Low
Bluetooth	Median	Shortest	Low	Phone accessories	Free	Low
ZigBee	Slow	Shorter	Low	Indoor equipment	Free	Low

build a straight and basic overview of the LoRa. It has a key role in establishing and maintaining costs. In comparison to other comparable technologies, the cost of communication is very negligible. Therefore, the cost from LoRa nodes to the gateway and the nube may be called free of charge.

12.1.1.1 LoRaWAN Architecture

LoRa is a methodology of spectrum spread modeling (CSS) that derives from the technology of chirp spread spectrum. LoRaWAN is the architecture of the LoRa Alliance's network system and communication protocol standard (Saari et al., 2018). LoRaWAN is a layered protocol for MAC, which allows LoRa to operate in larger applications. The LoRa Alliance specifies the network layer and network system architecture (see note 1). The various components connect end nodes to the cloud and the application servers in a LoRaWAN communication flow (Mekki et al., 2019).

12.2 LORAWAN COVERAGE IN REAL-TIME TEST SCENARIOS

The unique qualities of LoRa, particularly its low sensitivity to interference, minimal path loss, and excellent building penetration, allow for extremely long ranges even in densely populated areas. Even with modern software-assisted modeling approaches, however, the total range and coverage of a radio network can only be anticipated to a limited extent – and this is true for any radio technology, not only LoRa (Kanakaraja and Kotamraju, 2021). As a result, range and penetration testing are critical for better understanding radio wave behavior in real-world environments, particularly in the first phase.

Three documented tests in various settings are discussed in this chapter, as follows:

- Five distinct test situations in an urban environment.
- LoRaWAN in the suburban areas.
- Signal strength calculations inside basement rooms.

Only a few experimental range tests are adequately recorded in the literature in general.

12.2.1 TEST SCENARIO 1: SETTING OF LoRaWAN IN THE CENTER OF VIJAYAWADA (INDIA) CITY

Many of the cities have similar kind of architectural characteristics like Vijayawada (India) city in Andhra Pradesh, which we selected for our real-time testing for LoRaWAN coverage. There are just a few very tall structures, towering structures, and many that are only a few storeys high. These are not megacities with virtually entirely high-rise structures in the city core (Akram et al., 2021).

We could get coverage of well over 10km with exposed gateway sites in over 20 range tests in metropolitan situations. Individual measures will be covered in the sections that follow. Description of the real-time test setup (Ashok and Saleem Akram, 2020):

- These were non-line-of-sight (NLOS) connections in metropolitan environments.
- 12th spreading factor (maximum range).
- Maximum output power: 14 dBm.
- Frequency of the channel: 868 MHz band.
- The receiving gateway Rx was placed on the roof at the height of about 12 m, and the transmitter unit (Tx) was moved between five control stations.
- About 50 transmission attempts were made at each site.

In the following list, we have explain the features of the control points and the measurement series results:

Test Point 1: the signal passes across four buildings. A line of sight is not attainable along this path due to three high-rise structures and one low-rise structure. However, a direct channel to the receiver is also accessible. The distance between the two sites is 830 m. Ninety-six percent of the packets transmitted were received (Gopal and Akram, 2019).

Test Point 2: the signal passes across 14 buildings. Test Point 2 in this scenario is home to a massive concentration of low-rise residential complexes. There is also a block of apartments on the path to the receiver. The range is 960 m. Ninety-two percent of the data packets sent were received (Komalapati and Yarra, 2021).

Test Point 3: the signal passes across six buildings. This point is in the neighborhood on the side of a large plaza. The path passes through a large residential complex, followed by various industrial structures. The maximum range is 1,070 m. Once again, 98% of the packets sent were received.

Test Point 4: the signal passes through 14 buildings. It is the most significant way. This spans the apartment's four high-rise buildings. Then an open area with no obstacles follows until a grouping of high-rise skyscrapers reaches again. Finally, several industrial structures are located before the road to the recipient is built. The distance between us is 1,530 m. Ninety-eight percent of the packets transmitted were received.

Test point 5: the signal passes across six buildings. This test spot is divided into several industrial buildings that do not have open space between them. It is 863 m high. Received 100% of the transmission of data packets.

TABLE 12.2
Comparison of Various Test Points and Their Success Rates

Test Point	The Number of Buildings between Source to Destination	Distance to the Gateway (Rx) (m)	Success Rate (%)
1	4	830	96
2	14	960	92
3	6	1,070	98
4	14	1,530	98
5	6	863	100

Table 12.2 provides the (see note 1) locations of the test sites for the results of the successful tests in real time. All probable scenarios, such as overcrowded buildings at different distances, are covered by test points. The word "success rate" refers to the rate at which the data from a LoRa node to a gate and vice versa are successfully transmitted/received.

12.2.1.1 Assessment of Findings, Practical Observations, and Suggested Optimization Approaches

The LoRaWAN standard is the fact that not all packets are always received 100%, which were sent: the communication can be set, if required, as the recipient's message (so-called confirmed connections up and down), e.g., the sender who waits for a confirmation reception and who returns the data should not be given such an acknowledgment after a certain period. Therefore, all achieved levels may be regarded as noncritical, which is explained as follows:

- Three basic solutions for reach and coverage optimization are available.
- In typical LoRaWAN applications, broad coverage and ranges may be accomplished by allowing lower success rates, meaning accepting single failed signals and then retransmitting the ranges in the test might have been expanded.
- Redundant network topologies from several directions enable additional network coverage while preserving or enhancing success rates.
- The lighting on the surface might be substantially improved with a higher location of the doorway. Unfortunately, the door was only installed at 12 m in this test.
- The Hata-propagation model (Hata-propagation model) is a basic radio propagation model for calculating the loss of paths of radio waves outside. The initial range roughly doubles when the gateway's position is extended from 12 to 30 m, say.

12.2.2 SCENARIO 2: SETTING OF LoRaWAN IN THE SUBURBAN REGION

The entrance was located on the second floor of a building, directly before the window. At an external temperature of 15°C, the moisture content was 55% on the test day. The distance between the door and the five separate test stations was 650–3,400 m.

The terminal system was in a vehicle during the testing. The terminal device's transmission power is set at 14 dBm by default. The packet confirmation and retransmission were switched off to evaluate the performance of different spreading factors. However, the connection test was not disabled so that the propagation factor does not change even if a packet loss occurs. This is because the LoRaWAN network defaults to the connection quality in the distribution factor.

For these experiments were used the spreading factors 7, 9, and 12. Around 100 packets were broadcast to the network server with a sequence number for each test.

12.2.2.1 Test Results – Observations

The larger propagation factors have greater coverage than previously mentioned – more than 80% of the packets received at an SF value of 12 at 2,800 m and no packet at an SF of 7. At around 5 m above ground level, the gate was on the second storey. The test point was 2,800 m away, just behind a seven-storey building. SF12 alone at 3,400 m provided covermount. The price of the high rate using the high dispersion factor is a substantially lower bit rate. In contrast, network coverage with low dispersion factors is greatly reduced.

If the gateway is to be installed permanently, it should be placed in a more advantageous area. Unfortunately, the location is a building on the second storey, which is undoubtedly unsuitable for providing the widest coverage.

12.2.3 Scenario 3: In a Basement, LoRaWAN Is Used to Read the Water Meter, Gas Meter, and Electricity Meter

A scenario including cellar coverage is also stated in the previously referenced Libelium source. However, the basic criteria are only loosely described: In this measurement, the transmitter was positioned at an altitude of roughly 3 m outside the first floor of the office structure. The receiver was situated near the garage entrance below the ground floor. The point of view was not visible, and the range of LoRaWAN was 206 m. There are many more which are not documented.

The findings we achieved with consumers in penetration tests are far more illuminating. So we would like to give three more instances this time.

Data were provided to network a water meter during the first test of 998 m from a road ship to an entrance at the height of 35 m. Thus, a connection may be established despite the metal cover of the shaft. A delivery car was placed on the shaft to imitate the conditions in the real world, but signals were still in good form.

Penetration tests were performed at another client over many days in five different cellar rooms within a 4 km radius (Table 12.3).

12.2.3.1 Comprehensive Model 1: Indoor Coverage of LoRaWAN

Indoor and outdoor planning parameters are often taken into consideration while developing wireless network coverage. Statistical propagation models have therefore been built to assist the development of areas. For example, this is the Davide Magrin model (see network-level performances of a LoRa system). In addition, LPWAN communication networks, such as LoRaWAN, are essentially PMPs with sensors on different floors such as cellars. These requirements encourage network planners to build digital 3D maps and to employ deterministic distribution models.

Different approaches can be used for modeling losses generated from the outside and inside walls of buildings. For example, in one of the models offered, the following units evaluate the impacts of buildings on the route loss of LoRaWAN systems:

- Losses via building through exterior walls.
- Losses via building through inner walls.
- Higher energy use since one unit is above the first floor.

TABLE 12.3
Table of Success Rates Based on Distance and the Gateway to Which They Were Linked

	1		2		3		4		5	
	Distance (m)	Success rate (%)	Distance (m)	Success Rate (%)	Distance (m)	Success Rate (%)	Distance (m)	Success Rate (%)	Distance (m)	Success rate (%)
Gateway 1: 5 m	3,500	0	2,600	0	1,500	0	5,200	0	1,200	0
Gateway 2: 10 m	4,100	0	3,500	0	3,700	2	2,700	0	4,200	26
Gateway 3: 45 m	700	81	1,700	5	2,000	87	2,500	78	3,800	56

The following formula may be used to determine route losses through buildings using this model:

12.2.3.2 Building Losses via the Exterior Walls

The variability of materials and external wall thickness in many different buildings should be considered because the loss via external walls of both types of equipment is not usually the same. Table 12.4 illustrates three distinct ranges of values and the probability of a node losing this kind of value.

12.2.3.3 Losses Via Constructing Interior Walls

The effect of the inner walls is the highest value between the two values. Therefore, the number of internal barriers causing the loss is calculated as follows:

$$Gate\ 1 = Wi - p \tag{12.1}$$

where Wi is distributed in the [3, 10] dB range evenly and p is the internal wall count splitting the receiver from the transmitter.

It is assumed that 15% of the units are evenly distributed among the values of $p = \{0, 1, 2\}$ with a value of $p = 3$ for those units.

The second variable required to describe the loss of trajectory through inner barriers is calculated as follows:

$$Tor3 = \alpha d \tag{12.2}$$

where

$\alpha = 0.5$ dB/m – the penetration depth and the coefficient.
$d = $ a value that is equally scattered across the range of [0, 14] m.

12.2.3.4 Power Boost Received

The power signal contribution finally describes how improved the antenna height reception is achieved. The following expression defines this parameter:

$$Power\ signal = n * G_n \tag{12.3}$$

TABLE 12.4
A Probable Energy Signal Value Range and Likelihood

Probability	Ranger
0.24	[3, 10] dB
0.64	[10, 18] dB
0.1	[18, 22] dB

where

$Gn = 1.5\,\text{dB}$ – the increase owing to the height increase on one floor.
n = the number of buildings, divided equally throughout the values of $n = \{0\text{--}4\}$.

The three elements describe the overall indoor building loss are provided by the following equation:

$$L(\text{dB}) = \text{path loss} + \max\left[\text{Tor1}, \text{Tor}\right] - \text{GFH} \qquad (12.4)$$

In the virtual environment, certain tests with different scenarios suggest. However, no penetration could be ensured theoretically, calculating the complete loss on the direct route over multiple levels and walls. A gateway to the multistorey structure on the same roof with the basement or cellar to be connected may also have a sense.

There is rarely any importance to the straight passage through the building and reflection on nearby structures. The attenuation is inadequate to one to two walls, whereas the pathway is longer, and the structure reflexes.

12.3 CONCLUSIONS

This article presented the real-time test scenarios of the LoRaWAN nodes covered with different distances and various line-of-sight cases with practical demonstration and comparison of results in suburban and urban areas. Three types of cases are demonstrated as follows: (i) five various test cases – urban environment, (ii) LoRaWAN – suburban area, and (iii) the penetration of the basement rooms for reading data of the meter. Finally, we conclude that LoRaWAN nodes have a better coverage capacity with fewer losses, and the same is proven with test results and simulation plots. This is a practical demonstration verified in Vijayawada city in India.

REFERENCES

Akram, P.S., Ramesha, M., Ramana, T.V, Simulation and analysis of smart home architecture, *ARPN Journal of Engineering and Applied Sciences*, 2021, vol. 16, issue 6, pp. 280–286.

Ashok, G.L.P., Saleem Akram, P., Implementation of smart home by using packet tracer, *International Journal of Scientific and Technology Research*, 2020, vol. 9, issue 2, pp. 678–685.

Bourennane, E.-B., Boucherkha, S., Chikhi, S., A study of LoRaWAN protocol performance for IoT applications in smart agriculture, *Computer Communications*, 2020, vol. 164, pp. 148–157.

Butun, I., Pereira, N., Gidlund, M., *Analysis of LoRaWAN v1.1 Security: Research Paper*. 1–6, 2018. doi: 10.1145/3213299.3213304

de Carvalho Silva, J., Rodrigues, J.J.P.C., LoRaWAN —A low power WAN protocol for Internet of Things: A review and opportunities, *2017 2nd International Multidisciplinary Conference on Computer and Energy Science (SpliTech)*, 2017, pp. 1–6.

Gopal, S.R., Akram, P.S., Design, and analysis of heterogeneous hybrid topology for VLAN configuration, *International Journal of Emerging Trends in Engineering Research*, 2019, vol. 7, issue 11, pp. 487–491.

Kanakaraja, P., Kotamraju, S.K., IoT enabled ble and lora based indoor localization without GPS, *Turkish Journal of Physiotherapy and Rehabilitation*, 2021, vol. 32, issue 2, pp.642–653.

Komalapati, N., Yarra, V.C., Smart Fire Detection and Surveillance System Using IOT, *Proceedings - International Conference on Artificial Intelligence and Smart Systems, ICAIS 2021*, 2021, pp. 1386–1390.

Mekki, K., Bajic, E., Chaxel, F., Meyer, F., A comparative study of LPWAN technologies for large-scale IoT deployment, *ICT Express*, 2019, vol. 5, pp. 1–7. doi: 10.1016/j.icte.2017.12.005

Butun, I., Nuno Pereira, Mikael Gidlund. "Analysis of LoRaWAN v1. 1 security." In *Proceedings of the 4th ACM MobiHoc Workshop on Experiences with the Design and Implementation of Smart Objects*, pp. 1–6, 2018.

Sharan Sinha, R., Wei, Y., Hwang, S.-H., A survey on LPWA technology: LoRa and NB-IoT, *ICT Express*, 2017, vol. 3, issue 1, pp. 14–21.

NOTES

1 LoRa end nodes are not linked to a particular gateway, unlike cellular technology. Instead, data are delivered to numerous gateways through a node. Then each gateway passes the packet to a cloud-based LoRa network server (LNS) from the end node (cellular, ethernet, satellite, and WiFi). LNS is complicated and intelligent for network managing, filtering redundant packages, doing security checks, and allowing adaptive data rates (de Carvalho Silva and Rodrigues, 2017). Consequently, there is no transfer for a mobile LoRaWAN device from one gateway to another.

2 Proportional length to the gateway and the corresponding percentage of success. There have been thousands of tests of transmission spanning several hours or days at every site. Even at enormous distances, subterranean rooms with their impediments (reinforced cement, 2nd cellar, steel cellar doors, etc.). In this scenario, the height of the door is critical. In addition, this is based on situations requiring the retransmission of value if the success rate does not always guarantee direct transfer.

13 Intelligent Health Care Industry for Disease Detection

Ishaan Deep, Payaswini Singh, Aditi Trishna,
Sanskar, Biswa Ranjan Senapati,
and Rakesh Ranjan Swain
Siksha 'O' Anusandhan University

CONTENTS

13.1 INTRODUCTION

The state of one's health is crucial. "Health is Wealth," as the saying goes, because a healthy individual is eligible for a variety of benefits, including a decent job life and personal food habits. However, an ill individual is bound by numerous limitations. People's lifestyles are rapidly changing nowadays. This has an impact on their health. Life expectancy is dwindling, and health problems are on the rise.

Researchers are still looking into how to detect these diseases at an early age so that they can be cured easily. One of the research methods is the machine learning (ML) approach [1]. ML is a type of artificial learning (AL) in which various algorithms are used to determine the accuracy and precision values of an algorithm in order to determine whether it is producing the desired output [2]. ML is widely used in many fields to enhance performance. Various application domains in which ML models are widely used are healthcare [3–5], vehicular networking [6–11], cyber security [12], industry sector [13], etc. Figure 13.1 represents different types of ML.

This chapter focuses on the detection of diseases by means of ML algorithms, and so the authors have used supervised ML (support vector machine (SVM) and linear regression).

Supervised learning has one of the most popular algorithms known as SVM, which is used for classification as well as regression problems. In the SVM algorithm,

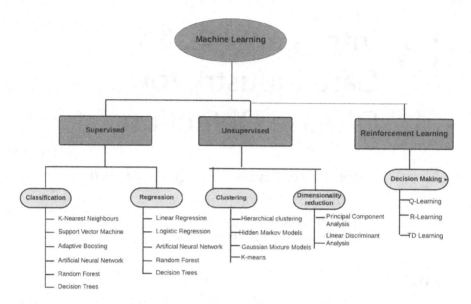

FIGURE 13.1   Different types of machine learning.

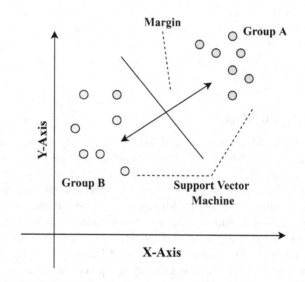

FIGURE 13.2   Support vector machine algorithm.

the data records are plotted as a position in a multidimensional space where every dimension has some features being the value of a particular coordinate. The SVM kernel is a functional approach for processing and converting low-dimensional input into high-dimensional space [14]. Figure 13.2 shows how SVM divides the dataset into two groups.

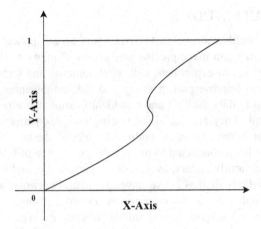

FIGURE 13.3 Logistic regression algorithm.

One of the most often used ML methods is logistic regression (LR), which is used to predict a categorical dependent variable using a set of independent variables. One of the most common models is a binary outcome. It is a useful method for analyzing problems and to fit the data into the best category [15]. The logistic function is depicted in Figure 13.3.

In this chapter, the authors have used datasets of three common diseases that the people are suffering from such as obesity, heart problem, and cancer. Nowadays, most of the people are suffering from these diseases at an early age, and thus the life span is decreasing. Using the ML algorithms, these diseases can be rectified at an early age and people can take the required precautions to live a healthy and happy life.

Death rates are also rising, with obesity, heart attacks, and cancer accounting for the majority of deaths. Because these diseases are hazardous, they should be recognized at a young age so that necessary safeguards can be taken and a happy and healthy life can be lived. The goal of this model is to figure out how to detect these diseases at a young age.

The major contribution of this chapter is presented as follows:

a. To detect the diseases at an early age using SVM and LR.
b. To compare the accuracy obtained for SVM and LR for the proposed work.
c. The performance is measured in terms of performance metrics like precision, recall, F1 score, and confusion matrix.

The following is the chapter's structure. The second section covers the literature study. The third section examines the proposed work. The fourth section presents the simulation results. The conclusion and future scope are discussed in the fifth section.

13.2   LITERATURE STUDY

When it comes to health informatics, ML has a very broad application. Its goal is to create algorithms that can make precise predictions. Primary health care is a vast field for ML in which to experiment with new domains and techniques. It offers a variety of tools to improve patient safety and risk management techniques. ML technology can aid in data analysis and risk identification, thereby improving diagnosis and treatment. They are useful for tracing complex patterns in large datasets. Cardiovascular disease refers to conditions affecting the heart or blood vessels. Many studies have been conducted to predict the prevalence of CVD using various factors such as age, family history, hypertension, cholesterol, and smoking. Usually, it is found that methods such as Naïve Bayesian, neural network, and decision tree algorithms were used. This chapter employs supervised learning models such as LR, decision tree, k-nearest neighbors, and random forest. Sex, type of chest pain, and number of major vessels are among the variables considered. The disease was correctly predicted 90.16% of the time [16].

Obesity is measured using BMI (or body mass index). The four ML techniques used in this chapter are binary LR, improved decision tree, weighted k-nearest neighbor, and artificial neural network. This study includes data from Tennessee high school students. Energy intake, physical activity, and sedentary behavior were among the variables studied. It is extremely useful because it defines the application of these ML techniques in great detail. It predicts obesity using two parameters: accuracy and specificity. The KNN model had the highest accuracy (88.82%), and the ANN model had the highest specificity (i.e., 99.46%) [17].

This chapter is based on the prediction of obesity using genetic variant analytics, where SNPs were used as variables in the prediction of obesity. Following feature selection, only 13 important variables were chosen. This chapter employs techniques such as k-nearest neighbor, SVM, and random forest. This chapter was extremely useful because it employed similar ML techniques. The authors made a significant observation: when the number of variables increases significantly, the performance of these ML models decreases. The chapter concludes that the SVM has the highest accuracy, 90.5% [18].

Cancer is one of the most dangerous diseases, and predicting it in its early stages is critical. Various methods have been used to detect this disease so that proper treatment can be provided before a patient's condition becomes too severe. There are numerous datasets available in the research community. As a result, accurately predicting the disease is both interesting and difficult. The prediction in this study is based on microarray datasets. It contains datasets on colon, leukemia, and lymphoma cancer. This study suggests that information gain and Pearson's correlation coefficient were the best feature selection methods for these three datasets, while MLP and KNN were the best classifiers [19].

13.3   PROPOSED WORK

In this model, the authors used three datasets—the obesity dataset [20], the heart attack dataset [21], and the cancer dataset [22]. The datasets were trained and tested

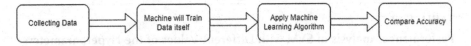

FIGURE 13.4 Working principle of machine learning model.

TABLE 13.1
Details of Baseline Methods of Machine Learning

State	Abbreviation
Artificial Learning	Al
Machine Learning	Ml
Support Vector Machine	SVM
Logistic Regression	LR
Cardiovascular Disease	CVD
Body Mass Index	BMI
K-Nearest Neighbor	KNN
Artificial Neural Network	ANN
Single Nucleotide Polymorphism	SNPs
Multilayer Perceptron	MLP

properly by using LR and SVM algorithms, where 90% of the dataset has been trained and 10% of the dataset has been tested to find its accuracy. The recall value, precision value, and the F1 score were also calculated. For three datasets, the confusion matrix has also been calculated using LR and SVM algorithms.

Figure 13.4 shows the working information of ML algorithms. The considered algorithm use for analysis is tabulated in Table 13.1

The abbreviation used here are given below:

13.4 SIMULATION RESULTS

The simulation results for the three datasets—obesity, heart attack, and cancer dataset—to find the accuracy, recall, precision, F1 score, and confusion matrix using SVM and LR are provided below, where C and Gamma are hypermeters. The training model is set using C and Gamma, where C regulates the error and Gamma determines the curvature weight of the decision border.

Performance metrics is based on four performance values, as follows:

a. **Accuracy:** it is the proportion between the number of accurate forecasts and all predictions combined [23].

$$\text{Accuracy} = \frac{\text{Number of correct predictions}}{\text{Total number of predictions}}$$

TABLE 13.2

Performance Analysis of SVM with Different Values of the Hyperparameters

	SVM			LR
	C	Gamma	Accuracy	Accuracy
Obesity	2	0.001	0.92	0.82
Heart attack	2	0.001	0.82	0.82
Cancer	2	0.0001	0.96	1

TABLE 13.3

The Comparative Analysis of Considered Performance Parameters

	SVM/LR		
	Recall	Precision	F1 Score
Obesity	0.9218/0.8224	0.9262/0.8252	0.9242/0.8228
Heart attack	0.8285/0.8280	0.8473/0.8218	0.8239/0.8244
Cancer	0.9588/1	0.9621/1	0.9597/1

b. **Precision:** it is calculated by dividing the total number of accurate forecasts by the number of true positives [24].

$$Precision = \frac{True\,positive}{True\,positive + False\,Negative}$$

c. **Recall:** it is the proportion of true positives to the sum of true positives and false negatives [24].

$$Precision = \frac{True\,positive}{True\,positive + False\,negative}$$

d. **F1 Score:** it refers to how many correct hits were also discovered, or how many true positives were recalled [24].

$$F1\,Score = \frac{2*Recall*Precision}{Recall*Precision}$$

13.5 CONFUSION MATRIX

To analyse the performance, the confusion matrix for Obesity using SVM and Logistic Regression has been drawn in Figure 13.5 and Confusion matrix for Heart Attack using SVM and Logistic Regression has been created in Figure 13.6, respectively.

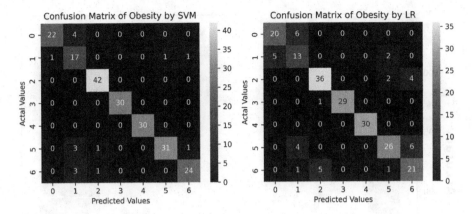

FIGURE 13.5 Confusion matrix for obesity using SVM and logistic regression.

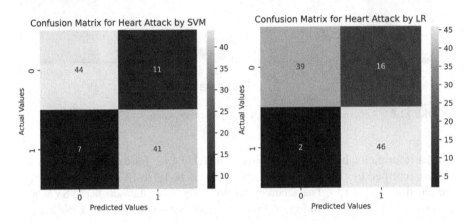

FIGURE 13.6 Confusion matrix for heart attack using SVM and logistic regression.

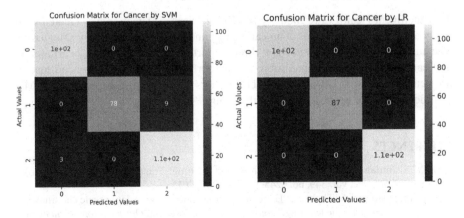

FIGURE 13.7 Confusion matrix for cancer using SVM and logistic regression.

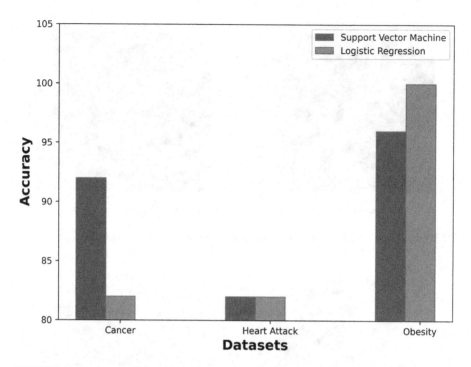

FIGURE 13.8 Accuracy graph for SVM and logistic regression.

Similar information has also been tabulated in Table 13.2. Further, the performance metrics returned by the SVM have been tabulated in Table 13.2. The same is also presented in Figure 13.7. The accuracy results of the three datasets using SVM and LR algorithms are shown in Figure 13.8. The same is also tabulated in Table 13.3.

13.6   CONCLUSION AND FUTURE SCOPE

The Intelligent Health Care Industry Model for Disease Detection was created with the goal of detecting diseases at an early stage and taking the necessary safeguards. The authors have taken into account three important diseases: obesity, heart attack, and cancer. The accuracy, precision, recall score, F1 score, and confusion matrix are determined using LR and SVM. The authors will aim to incorporate other ML techniques into the proposed work in the future.

REFERENCES

1. M. I. Jordan and T. M. Mitchell, "Machine learning: Trends, perspectives, and prospects," *Science*, vol. 349, no. 6245, pp. 255–260, 2015.
2. P. Prajapati and A. Thakkar, "Extreme multi-label learning: A large scale classification approach in machine learning," *Journal of Information and Optimization Sciences*, vol. 40, no. 4, pp. 983–1001, 2019.

3. A. Kumar, R. Kumar, and S. S. Sodhi, "Intelligent privacy preservation electronic health record framework using soft computing," *Journal of Information and Optimization Sciences*, vol. 41, no. 7, pp. 1615–1632, 2020.
4. A. Vellido, "The importance of interpretability and visualization in machine learning for applications in medicine and health care," *Neural Computing and Applications*, vol. 32, no. 24, pp. 18069–18083, 2020.
5. J. Wiens and E. S. Shenoy, "Machine learning for healthcare: On the verge of a major shift in healthcare epidemiology," *Clinical Infectious Diseases*, vol. 66, no. 1, pp. 149–153, 2018.
6. B. R. Senapati, P. M. Khilar, and R. R. Swain, "Fire controlling under uncertainty in urban region using smart vehicular ad hoc network," *Wireless Personal Communications*, vol. 116, no. 3, pp. 2049–2069, 2021.
7. B. R. Senapati, P. M. Khilar, and R. R. Swain, "Composite fault diagnosis methodology for urban vehicular ad hoc network," *Vehicular Communications*, vol. 29, p. 100337, 2021.
8. B. R. Senapati, P. M. Khilar, and R. R. Swain, "Environmental monitoring through vehicular ad hoc network: A productive application for smart cities," *International Journal of Communication Systems*, vol. 34, no. 18, p. e4988, 2021.
9. B. R. Senapati and P. M. Khilar, "Optimization of performance parameter for vehicular ad-hoc network (vanet) using swarm intelligence," *Nature Inspired Computing for Data Science*, 2020, pp. 83–107.
10. B. R. Senapati, P. M. Khilar, T. Dash, and R. R. Swain, "Automatic parking service through VNET: A convenience application," in *Progress in Computing, Analytics and Networking: Proceedings of ICCAN*, 2019 (pp. 151–159), Springer Singapore, 2020.
11. B. R. Senapati and P. M. Khilar, "Vehicular network based emergency data transmission and classification for health care system using support vector machine", 2022.
12. C. Virmani, T. Choudhary, A. Pillai, and M. Rani, "Applications of machine learning in cyber security," in P. Ganapathi and D. Shanmugapriya (Eds), *Handbook of Research on Machine and Deep Learning Applications for Cyber Security*. IGI Global, 2020, pp. 83–103.
13. D. P. Penumuru, S. Muthuswamy, and P. Karumbu, "Identification and classification of materials using machine vision and machine learning in the context of industry 4.0," *Journal of Intelligent Manufacturing*, vol. 31, no. 5, pp. 1229–1241, 2020.
14. S. Suthaharan, "Support vector machine," in S. Suthaharan (Ed.) *Machine Learning Models and Algorithms for Big Data Classification: Thinking with Examples for Effective Learning*. Springer, 2016, pp. 207–235.
15. D. W. Hosmer Jr, S. Lemeshow, and R. X. Sturdivant, *Applied Logistic Regression*. John Wiley & Sons, 2013, vol. 398.
16. M. Nabeel, M. J. Awan, M. Raza, H. Muslih-Ud-Din, and S. Majeed (2021, November). Heart attack disease data analytics and machine learning. In *2021 International Conference on Innovative Computing (ICIC)* (pp. 1–6). IEEE.
17. Z. Zheng, and K. Ruggiero (2017, November). Using machine learning to predict obesity in high school students. In *2017 IEEE International Conference on Bioinformatics and Biomedicine (BIBM)* (pp. 2132–2138). IEEE.
18. C. A. C. Montañez, P. Fergus, A. Hussain, D. Al-Jumeily, B. Abdulaimma, J. Hind, and N. Radi (2017, May). Machine learning approaches for the prediction of obesity using publicly available genetic profiles. In *2017 International Joint Conference on Neural Networks (IJCNN)* (pp. 2743–2750). IEEE.
19. S. B. Cho, and H. H. Won (2003, January). Machine learning in DNA microarray analysis for cancer classification. In *Proceedings of the First Asia-Pacific Bioinformatics Conference on Bioinformatics 2003-Volume 19* (pp. 189–198).
20. https://www.kaggle.com/datasets/ankurbajaj9/obesity-levels?select=ObesityDataSet_raw_and_data_sinthetic.csv

21. https://www.kaggle.com/datasets/johnsmith88/heart-disease-dataset
22. https://www.kaggle.com/datasets/rishidamarla/cancer-patients-data
23. https://developers.google.com/machine-learning/crash-course/classification/accuracy#:~:text=Accuracy%20is%20one%20metric%20for, predictions%20Total%20number%20of%20predictions
24. https://towardsdatascience.com/essential-things-you-need-to-know-about-f1-score-dbd973bf1a3#:~:text=1.-, Introduction, competing%20metrics%20E2%80%94%20precision%20and%20recall

14 Challenges with Industry 4.0 Security

Mohit Sajwan and Simranjit Singh
Bennett University Greater Noida

CONTENTS

14.1 INTRODUCTION

The most significant turning points in human history have been the industrial revolutions. Many researchers believe the industrial revolution has greatly impacted people's daily lives more than the scientific revolutions. A substantial shift has occurred in the sector over the past few decades as new developments have been integrated into the system. Industry 4.0 (also known as the "fourth industrial revolution") is a significant outcome of the digitalization of a sector characterized by inflexible standards and outdated fashions [1]. Manufacturing facilities have made significant

DOI: 10.1201/9781003321149-14

investments in developing intelligent production systems, which is a vital step toward Industry 4.0. Industries are now looking to integrate technology and techniques such as IoT, cloud solutions, and machine learning (ML) in order to improve productivity and reduce costs [2].

Throughout history, revolutions have developed rapidly in response to the changing requirements of each generation. Electronic automation in industrial environments began in the 1970s, and the Internet of Things (IoT) has recently evolved with data-driven production systems based on the cyber-physical system (CPS) [3]. An attempt to make German industry more competitive by modernizing it was the driving force behind the coining of the term "Industry 4.0." The word "fourth industrial revolution" began to be used interchangeably with this phenomenon as it spread over the world. This industry's rapid growth means that businesses will need to implement a new process management strategy. The procedure itself needs to be carried out as the primary objective. So, networked machines can change their production processes on their own based on criteria that have been set up in advance and without making any mistakes [4].

Security and privacy concerns are a result of the fourth industrial revolution and some of its major characteristics. Our current manufacturing and production industry is undergoing a digital change known as "Industry 4.0," as previously stated. Artificial intelligence, CPSs, cloud computing, and the IoT are just a few of the vital factors of Industry 4.0. Because factory systems are designed to run on all 7 days of the week, there can be no extended outages due to an attack. Intelligent manufacturing firms are beginning to see the value in knowing and defending their systems, as well as interpreting the data they generate [5].

In the context of Industry 4.0, it is absolutely necessary to devise an innovative cyber security strategy and approach. The security operations center (SOC) is in charge of combating these dangers, and it does so by constantly monitoring and controlling the equipment under its control. Threat detection and response, along with the speedy discovery of security concerns, are the primary functions of a SOC. We can lessen the damage of an attack if we can find threats early and let local teams know so they can fix security problems quickly without stopping production [6]. If IoT grows quickly by 2022, attackers will target it more [7].

Many corporations and organizations predict an exponential rise in Internet-connected devices in the future [8]. Gartner estimates 50 billion IoT connections for Cisco by 2020, 20.8 billion for Gartner, and 100 billion for Huawei by 2025. The most important lesson is to expect strong growth despite wide estimates. The number of Internet-connected devices will expand rapidly, requiring a comprehensive security solution [9]. Figure 14.1 shows Internet-connected device adoption from 2012 to 2025.

In recent years, cyber attacks have become more widespread, with victims ranging from people to entire countries. The year 2014 was dubbed the Year of the Breach, whereas 2015 was dubbed the Year of the Breach 2.0 by several industry experts. It is clear that cyber attacks have wreaked havoc over the world from a global perspective. Customers who use the IoT systems should be taught how to protect themselves against cyber attacks by their employers [10].

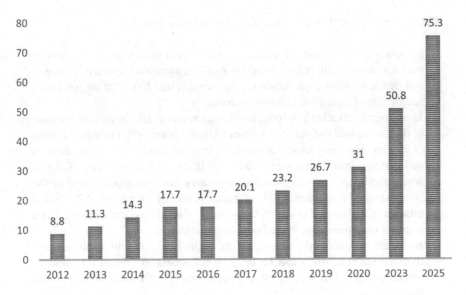

FIGURE 14.1 Connected IoT devices worldwide from 2012 to 2025.

Some thoughts on the problems of Industry 4.0, focusing on security, are presented in this paper in order to increase awareness of the need for security in Industry 4.0. The following is the outline for this document: After providing an introduction of manufacturing industry procedures in Section 14.1, Section 14.2 provides an overview of the security and vulnerability concerns over industry 4.0, after which Section 14.3 provides security challenges of the industry 4.0, and Section 14.4 concludes with conclusions and future developments.

14.2   SECURITY THREATS AND VULNERABILITIES

14.2.1   Cyber-Physical Systems

14.2.1.1   Cyber-Physical System Security Threats

When a cyber-physical threat is paired with one of the other risks that are listed below, it is a CPS security threat.

14.2.1.2   Cyber Threats

To properly center information, you must secure the flow of data not only while it is being stored but also while it is being sent and even while it is being processed [11].

Oriented Function: In order to do this, the cyber-physical components of the whole CPS need to be integrated.

Oriented Threat: A threat that has an influence on the data's availability, integrity, confidentiality, and accountability.

Due to the aforementioned problems, CPSs are susceptible to

1. System wireless capabilities can be used to gain remote access or control over a system or interrupt its operations if the system's structure is known, and hence wireless exploitation is necessary. Wired Exploitation Accidents and/or loss of control are likely outcomes.
2. Jamming: the attacker's purpose in this scenario is usually to modify the state of the device and the planned actions to inflict harm [12]. Deauthentication or wireless jamming signals are sent out in waves, and this prevents devices and systems from receiving services. An illustration of this type of risk is when intelligence services consistently carry out operations aimed at the computational intelligence (CI) of a nation, mostly by means of the transmission of malware. Therefore, the secrecy of the data is compromised as a result of the limitations of traditional protections.
3. An attack on a network, either a logical or physical one, can result in unauthorized access, which can be used to get access to critical data and so breach privacy. Physical or logical intrusions into the network can cause this.
4. Interception: hackers can listen in on private conversations by exploiting weaknesses that already exist or creating new ones, which results in yet another breach of privacy and confidentiality [13].
5. Hackers can monitor a device or even a vehicle by taking advantage of global positioning system (GPS) navigation systems, which violates users' expectations of having their privacy regarding their location protected.
6. Information Gathering: software makers stealthily collect audit logs and files that are stored on each device to get information they can sell for marketing and business purposes. They do this so they can sell a lot of personal information in an unethical way.

14.2.1.3 Physical Threats

Industrial neighborhood area networks (NANs), advanced metering infrastructure (AMI), and data meter management systems are being introduced to industrial CPSs to preserve their resilience. Physical risks can be categorized into three elements [12].

Damage from Physical Threats: power-generating stations (i.e., the base stations, power plants, and power grid) are exceptionally well guarded since different types of facilities employ diverse degrees of security. Authentication, authorization, and access control methods, including video surveillance, biometrics, access cards, and usernames and passwords, have been implemented at these stations, so they have enough staff and security. As a result of their high levels of security and staffing, these facilities have been able to remain open. However, the primary source of concern is the power generation substations, which have fewer robust physical safeguards in place [13].

The possibility of a hostile adversary bringing down more than one substation at the same time is the most worrisome prospect of the loss. In the event that there is significant damage to the smart grid, large metropolitan regions may experience a blackout that lasts for a number of hours. The People's Liberation Army (PLA), a Chinese organization with political motivations, was responsible for the cascading blackout that occurred in the United States on August 14, 2003 [14]. This incident is an example of a real-life worst-case scenario.

Repair: if a problem is detected, it may either be isolated and fixed manually, or an alarm may be sent to the associated control system to automatically rearrange the backup resources so that the service can be provided at all times. The self-healing method, which relies on the system's ability to recognize and isolate errors and interruptions, can be used to cure the problem, or it can be sent to a professional for repair. The goal here is to expedite the patient's recuperation in the shortest amount of time possible. However, critical components sometimes have no backup capacity at all, or only a limited amount of it. Therefore, self-healing is better able to deal with more serious injuries in a shorter period of time [15].

14.2.1.4 Vulnerabilities

There are three major categories of CPS vulnerabilities [16], as follows:

1. Platform: hardware, software, configuration, and database faults all fall under the umbrella of "platform vulnerabilities" (PV). There are many types of network vulnerabilities, but the most common are packet manipulation attacks, denial-of-service (DoS) and distributed denial-of-service (DDoS) attacks, back doors, communication-stack attacks, spoofing (on the application, transport, and network layer), sniffing, replay, eavesdropping, man-in-the-middle attacks, and others that compromise open wireless or wired communication and connections [17].
2. **Platform**: hardware, software, configuration, and database faults all fall under the umbrella of "PV".
3. **Management**: there are security rules, processes, and policies missing from the management structure [18].

14.2.2 INTERNET OF THINGS

14.2.2.1 Threats

An assault that is also known as a man-in-the-middle attack occurs when an adversary completely or partially modifies the data stream of an IoT device in order to impersonate the device or system from which the data originated. They are able to see data that is being transmitted, manipulate equipment, and intercept shared sensitive information [19].

Dangers associated with the sharing of information in order to gain information without proper authority, attackers, may listen in on broadcasts and eavesdrop, jam the signal to prevent information from being distributed, or partly override the broadcast in order to substitute it with fraudulent information [20]. They then threaten to make the information public or sell it.

Attackers have the ability to obtain access to the firmware or operating system of a device that is running an IoT application and subsequently changes it, either partially or entirely, on the device. Then, in order to access the network and any other associated services, they will utilize the authentic IDs of the device and application. Assaults such as SQL or XML injection, as well as distributed denial-of-service attacks, are examples of tampering concerns for the IoT applications [21].

Threats posed by the elevation of privilege attackers create damage by modifying the access control rules of an application by using unprotected IoT applications. For instance, in a manufacturing or industrial setting, an attacker may do harm to the production system or personnel by forcing a valve to open all the way when it should only open halfway. This would be contrary to the intended behavior of the valve.

Malware is still capable of infecting the majority of devices that are connected to the IoT [22], despite the fact that these devices have little processing capability. In the most recent few years, fraudsters have been very successful when making use of this tactic. IoT botnet malware is one of the most common types since it is both adaptable and profitable for cybercriminals. It is also one of the most often encountered forms. The most prominent cyberattack occurred in 2016 when Mirai brought major websites and online services to a halt by enlisting the help of a legion of common IoT devices. Malware that mines cryptocurrencies and ransomware are two more categories of malicious software.

An increase in the number of cyberattacks. DDoS attacks, often known as DDoS assaults, frequently make use of infected devices [23]. Devices that have been compromised can also be used as an attack base to infect further machines and disguise malicious behavior. They can also serve as an entry point for lateral movement within a corporate network. Even while businesses would appear to be the more lucrative targets, even smart homes are subject to a startling amount of cyberattacks that were not anticipated.

Theft of information and possible disclosure to third parties. Connected gadgets, like anything else that has to do with the internet, enhance the likelihood of exposure to threats on the internet. Important information, both technical and personal, might be mistakenly kept in these devices and used to target individuals [22].

Poor management and incorrect setting of the device. Inadequate management of devices, carelessness with passwords, and other security oversights can all contribute to the success of these assaults. There is also the possibility that users just lack the knowledge and competence to adopt appropriate security measures, in which case service providers and manufacturers may need to assist their clients in achieving greater protection.

14.2.2.2 Vulnerabilities

IoT applications suffer from various vulnerabilities that put them at risk of being compromised [24] are as follows:

1. Passwords that are easy to guess or are hard coded. There are a lot of passwords that are easy to guess, exposed to the public, or cannot be changed. Some members of the IT team do not bother changing the device's or software's factory-default password once they install it.
2. The absence of a system or procedure for updating. Many IoT applications and devices do not receive updates because the administrators of IT systems cannot see them on the network. Additionally, IoT devices may not even

have an update mechanism built into them owing to their age or function, so administrators cannot frequently update the software on these devices.

3. Network services and ecosystem interfaces that are not properly safeguarded. Every connection made by an IoT app has the potential to be hacked, either as a result of a flaw that is built into the components themselves or as a result of the fact that they are not protected from intrusion. This encompasses any gateway, router, modem, external web app, application programming interface (API), or cloud service connected to an IoT app.

4. IoT app components are either obsolete or not secure. When being constructed, many applications for the IoT make use of frameworks and libraries provided by third parties. They may be a threat to network security if they are out of date, have been found to be vulnerable in the past, and their installation in a network is not confirmed [23, 24].

5. Storing and transmitting data in an unsecured environment. Between the IoT apps and other connected devices and systems, it is possible to store and transfer many different sorts of data. Everything needs to be encrypted and securely protected using transport layer security or one of the other protocols, depending on the situation.

14.2.3 CLOUD COMPUTING

14.2.3.1 Threats

Common threats to cloud data security are as follows:

1. **Identity, Authentication, and Access Management**: this includes the absence of multifactor authentication, incorrectly set access points, weak passwords, scalable identity management systems, and continual automated rotation of cryptographic keys, passwords, and certificates. There is also a lack of continual automatic rotation of cryptographic keys and passwords.

2. **Vulnerable public APIs**: The development of application programming interfaces is necessary to prevent sensitive data from being accessed maliciously as well as accidentally. In addition to authentication and access control, it is recommended that encryption be used, as should the ability to keep track of user activities.

3. The attackers may attempt to spy on user activities and transactions, change data, return fraudulent information, and redirect users to sites that are not legitimate.

4. Malicious insiders. It is possible for a current or former employee or contractor of an organization with authorized access to the company's network, systems, or data to willfully misuse the access in a manner that leads to a data breach or negatively impacts the organization's information systems. Insider threats include malicious insiders.

5. Sharing data is a kind of cooperation. Data exchange between businesses are made simple by several cloud providers. As a result, fraudsters now have a larger target pool from which to launch attacks and steal critical data.

6. **Denial-of-service attacks**: As a result of cloud infrastructure disruption, hackers can inflict damage on enterprises without having access to their cloud service accounts or internal network.

14.2.3.2   Vulnerabilities

1. **Cloud storage with incorrect configuration**: data theft is made easier because of the widespread use of cloud storage. Businesses continue to configure cloud storage inappropriately despite the severe dangers, resulting in huge financial losses for many [24]. **Inadequate constraints or protections**: unauthorized access to your cloud infrastructure might put your organization at risk if you do not have enough controls or protections in place. Insecure cloud storage buckets can lead to attackers gaining access to your company's secret data and downloading it, which can have serious consequences. The default setup of Amazon Web Services (AWS) led to a huge number of data breaches since the S3 buckets were accessible by default [25].

2. **Theft or loss of an organization's intellectual property**: intellectual property (IP) is unquestionably one of the most precious assets that a business possesses; but it is also extremely susceptible to security risks, particularly when the data in question is kept online [26]. Cloud-based file-sharing systems have been found to include sensitive information such as intellectual property in roughly 21% of the files uploaded. When cloud services are breached, attackers may be able to access sensitive information stored in them.

   Changes made to the data: if the data are changed in a way that prevents them from being returned to their original condition, this might lead to a loss of the data's entire integrity and could render the data unusable.

   Deletion of Data: An adversary's ability to remove critical data from a cloud service, which constitutes a clear and present danger to an organization's operations, is a major concern when it comes to data security.

   Loss of Access: attackers can lock up information and demand a payment in order to release it (this is known as a ransomware attack), or they can encrypt data using robust encryption keys until they carry out their nefarious operations. As a result, operating in the cloud necessitates putting in place precautionary measures to secure your intellectual property and data [17].

3. **Unsatisfactory Management of Access**: incorrect management of user access is perhaps the most prevalent security concern associated with cloud computing. In security breaches involving online applications, credentials that have been stolen or lost have been the instrument that has been utilized by attackers the most frequently for the past many years.

   Access management guarantees that persons are only able to carry out the activities for which they are responsible. Authorization refers to the process of determining what a person is permitted to access and validating that permission. In addition to the standard access management issues that plague organizations in the modern day, such as user password fatigue and managing a distributed workforce, there are also a number of cloud-specific challenges that organizations must contend with. These include the following issues and more [17, 23]:

    a. Users who are allocated but are not active.

    b. Multiple administrator accounts.

    c. Users avoiding business access management controls.

4. **Devious Actors on the Inside**: even if you take precautions to protect yourself from the various potential threats posed by cloud computing, you may still be susceptible to assaults by malevolent insiders, including both current and past employees.

    If your adversaries are able to get illegal access to your systems, they have the capability of stealing information, destroying data, and sabotaging your IT infrastructure. Research published by Ponemon in 2020 states that the number of insider attacks has grown by 47% since 2018. The cost of insider attacks has increased by 31% since 2018. Only 23% of insider threats are intentional; the majority of insider attacks are due to incompetence.

5. **Insecure APIs**: the usage of application programming interfaces, or APIs, is a common way to make cloud computing more efficient. APIs, or application programming interfaces, simplify the process of exchanging data across many software programs and are therefore widely utilized in workplaces. APIs, which are well known for their practicality and the capacity to improve efficiency, can also be a source of vulnerabilities in the cloud [13].

    Attackers can readily access company data and conduct DDoS assaults by exploiting weak APIs. When they initiate attacks against APIs, sophisticated attackers might employ a number of techniques to avoid being discovered by security systems.

    As a result of the rising reliance that businesses have on APIs, there has been a rise in the number of attacks that target APIs. Abuse of application programming interfaces is projected to become the most prevalent type of attack vector by the year 2022, as stated by Gartner.

14.3   SECURITY CHALLENGES OF THE INDUSTRY 4.0

14.3.1   INADEQUATE IT/OT SECURITY EXPERTISE, KNOWLEDGE, OR UNDERSTANDING

As mentioned above, this difficulty is caused by personnel participating in changing industrial processes or digital security. Both lack information security skills. People involved in the manufacturing process do not know what security measures to take, and those in charge of security do not fully comprehend the process to secure it from outside threats. Cybersecurity training are getting fewer and more expensive [27].

14.3.2   INADEQUATE SECURITY POLICIES AND FUNDS

The 4th industrial revolution sector is still unwilling to focus on security. The lack sees this of policies and organizational standards that assure product privacy and security. No matter how much we stress it, the industry has not embraced privacy and security. Most operators consider it an add-on or luxury function, which leads to our following argument [28].

For the same reason that industries and operators fail to prioritize security, they have cut back on money and are reluctant to engage in R&D for current security concerns. This creates problems.

A corporation that migrates its information systems to the cloud instead of storing them locally isn't spending enough on security. Now, when a corporation discovers how much it is wasting on saving money by migrating to cloud services is a major motivator for storing data locally. How much would a corporation spend on cloud security if it only wanted to save money [29]? Cybersecurity will not improve as it should.

14.3.3 THE LEGAL OBLIGATION REGARDING PRODUCTS

Next, industry 4.0 operators have a security challenge with procedures. The problem is Industry 4.0 cybersecurity liability. Many stakeholders are engaged in making a smart device that can be connected to the internet via IoT. These stakeholders include the manufacturing brand, companies who supply one of the numerous parts used to make that smart device, security teams working on the security measures included in that smart device, software teams that program that device, and many more in the supply chain. This large number of stakeholders apportions accountability in case a smart device security event occurs, which might be a big obstacle in implementing industry 4.0 [30].

14.3.4 UNEVEN AND INSUFFICIENT STANDARDIZATION

Over the ages, science has advanced greatly. Thousands of studies, publications, and standards make these technologies easy to deal with, but not Industry 4.0 security. Unlike IoT or other new technologies with comprehensive standardization, Industry 4.0 security has little or no market standards. Only fragmented or unagreed-upon standards are used by existing Industry 4.0 operators [31].

14.3.5 THE DEVICES' CAPABILITIES AND LIMITATIONS FROM A TECHNICAL PERSPECTIVE

The digitalization of existing industrial processes is a key component of industry 4.0. Thus it is not a brand-new system. Instead, the identical tasks previously performed by hand will now be carried out digitally. This means that digital platforms will be incorporated into existing platforms, which is the only way to avoid starting from scratch. This easy technique to adopt Industry 4.0 has several drawbacks. Most gadgets now in use or made using obsolete approaches have technological limitations that pose security risks. Most of these devices have low power and can't do simple tasks. One of the main reasons is to keep pricing stable. All these already-operating devices lack the foundation to implement any protective measures. All this equipment was intended for one major operation, yet their protection was ignored. Many new devices lack this basic protection, too [6, 18].

14.4 CONCLUSION

In today's manufacturing industry, moving toward Industry 4.0 presents a wide range of technological difficulties, many of which have a significant impact on the security domain. An incident that has substantial and dramatic ramifications for the organization is normally handled after the development process is complete. Organizational security is commonly defined as a financial investment with no return on investment, which is why this is the case Return on Investment (ROI). Although this procedure is expensive, it often does not provide a long-term solution to the problem at hand. This may also impair the source of differentiation among rivals and diminish the competitive advantages and organizational trust in their commercial activities. Prior to beginning the implementation of Industry 4.0's technological problems, it is vital to design a strategy that includes all stakeholders and to agree on security issues as well as the appropriate architecture. Security concerns are already being brought up. Concerns about security and putting in place suitable organizational safeguards to protect processing activities have been reinforced by European Union (EU) directive 2016/679. Considering how important it is to secure the IT systems in industrial equipment, these talks will make the next stage of industrial development safer and more secure for everyone.

The experts in the working groups will play a crucial role in highlighting the various options that organizations will have in the near future. Even so, it is important to find and compare solutions that have worked well in the past and can help all stakeholders in the different business sectors.

REFERENCES

1. Mohamed Abomhara and Geir M. Køien. Cyber security and the internet of things: Vulnerabilities, threats, intruders and attacks. *Journal of Cyber Security and Mobility*, 4: 65–88, 2015.
2. Nilufer Tuptuk and Stephen Hailes. Security of smart manufacturing systems. *Journal of Manufacturing Systems*, 47: 93–106, 2018.
3. Beyzanur Cayir Ervural and Bilal Ervural. Overview of cyber security in the industry 4.0 era. In *Industry 4.0: Managing the Digital Transformation*, pages 267–284. Springer, 2018.
4. T. Pereira, L. Barreto, and A. Amaral. Network and information security challenges within industry 4.0 paradigm. *Procedia Manufacturing*, 13: 1253–1260, 2017.
5. Katalin Ferencz, József Domokos, and Levente Kovács. Review of industry 4.0 security challenges. In *2021 IEEE 15th International Symposium on Applied Computational Intelligence and Informatics (SACI)*, pages 245–248. IEEE, 2021.
6. Mohammed M. Alani and Mohamed Alloghani. Security challenges in the industry 4.0 era. In *Industry 4.0 and Engineering for a Sustainable Future*, pages 117–136. Springer, 2019.
7. Capgemini (2015). Securing the internet of things opportunity: Putting cybersecurity at the heart of the IoT. *Capgemini Worldwide*. Retrieved March 10, 2021, from https://www.capgemini.com/resources/securing-the-internet-of-things-opportunity-putting-cybersecurity-at-the-heart-of-the-iot/.

8. Rolf H. Weber and Evelyne Studer. Cybersecurity in the internet of things: Legal aspects. *Computer Law & Security Review*, 32(5): 715–728, 2016.
9. Abid Haleem, Mohd Javaid, Ravi Pratap Singh, Shanay Rab, and Rajiv Suman. Perspectives of cybersecurity for ameliorative industry 4.0 era: A review-based framework. *Industrial Robot: The International Journal of Robotics Research and Application*, 49: 582–597, 2022.
10. Jean-Paul A. Yaacoub, Ola Salman, Hassan N. Noura, Nesrine Kaaniche, Ali Chehab, and Mohamad Malli. Cyber-physical systems security: Limitations, issues and future trends. *Microprocessors and Microsystems*, 77: 103201, 2020.
11. Michael Rushanan, Aviel D. Rubin, Denis Foo Kune, and Colleen M. Swanson. Sok: Security and privacy in implantable medical devices and body area networks. In *2014 IEEE Symposium on Security and Privacy*, pages 524–539. IEEE, 2014.
12. Stephen Checkoway, Damon McCoy, Brian Kantor, Danny Anderson, Hovav Shacham, Stefan Savage, Karl Koscher, Alexei Czeskis, Franziska Roesner, and Tadayoshi Kohno. Comprehensive experimental analyses of automotive attack surfaces. In *20th USENIX Security Symposium (USENIX Security 11)*, 2011.
13. Hossein Zeynal, Mostafa Eidiani, and Dariush Yazdanpanah. Intelligent substation automation systems for robust operation of smart grids. *2014 IEEE Innovative Smart Grid Technologies-Asia (ISGT ASIA)*, pages 786–790, 2014.
14. Thomas M. Chen, Juan Carlos Sanchez-Aarnoutse, and John Buford. Petri net modeling of cyber-physical attacks on smart grid. *IEEE Transactions on Smart Grid*, 2(4): 741–749, 2011.
15. A. Muir and J. Lopatto. Final report on the August 14, 2003 blackout in the United States and Canada: Causes and recommendations. 2004.
16. Hang Zhang, Bo Liu, and Hongyu Wu. Smart grid cyber-physical attack and defense: A review. *IEEE Access*, 9: 29641–29659, 2021.
17. Narendra Mishra, R.K. Singh, and Sumit Kumar Yadav. Design a new protocol for vulnerability detection in cloud computing security improvement. In *Proceedings of the International Conference on Innovative Computing & Communication (ICICC)*, 2021.
18. Jacob Morgan. A simple explanation of the internet of things. *Retrieved November*, 20: 2015, 2014.
19. Derui Ding, Qing-Long Han, Yang Xiang, Xiaohua Ge, and Xian-Ming Zhang. A survey on security control and attack detection for industrial cyber-physical systems. *Neurocomputing*, 275: 1674–1683, 2018.
20. Naser Hossein Motlagh, Mahsa Mohammadrezaei, Julian Hunt, and Behnam Zakeri. Internet of things (IoT) and the energy sector. *Energies*, 13(2): 494, 2020.
21. Zhiyan Chen, Jinxin Liu, Yu Shen, Murat Simsek, Burak Kantarci, Hussein T. Mouftah, and Petar Djukic. Machine learning-enabled IoT security: Open issues and challenges under advanced persistent threats. *ACM Computing Surveys (CSUR)*, 55(5): 1–37, 2022.
22. Arup Barua, Md Abdullah Al Alamin, Md Shohrab Hossain, and Ekram Hossain. Security and privacy threats for bluetooth low energy in IoT and wearable devices: A comprehensive survey. *IEEE Open Journal of the Communications Society*, 3: 251–281, 2022.
23. Ritika Raj Krishna, Aanchal Priyadarshini, Amitkumar V. Jha, Bhargav Appasani, Avireni Srinivasulu, and Nicu Bizon. State-of-the-art review on IoT threats and attacks: Taxonomy, challenges and solutions. *Sustainability*, 13(16): 9463, 2021.
24. Omar Alrawi, Charles Lever, Kevin Valakuzhy, Kevin Snow, Fabian Monrose, Manos Antonakakis, et al. The circle of life: A {Large-Scale} study of the {IoT} malware life-cycle. In *30th USENIX Security Symposium (USENIX Security 21)*, pages 3505–3522, 2021.

25. Manish Snehi and Abhinav Bhandari. Vulnerability retrospection of security solutions for software-defined cyber–physical system against DDoS and IoT-DDoS attacks. *Computer Science Review*, 40: 100371, 2021.
26. Kurdistan Ali and Shavan Askar. Security issues and vulnerability of IoT devices. *International Journal of Science and Business*, 5(3): 101–115, 2021.
27. Jan Svacina, Jackson Raffety, Connor Woodahl, Brooklynn Stone, Tomas Cerny, Miroslav Bures, Dongwan Shin, Karel Frajtak, and Pavel Tisnovsky. On vulnerability and security log analysis: A systematic literature review on recent trends. In *Proceedings of the International Conference on Research in Adaptive and Convergent Systems*, pages 175–180, 2020.
28. David Talbot. Vulnerability seen in amazon's cloud computing. *Technology Review*. Retrieved from http://www.cs.sunysb.edu/~sion/research/sion, 2009.
29. Sarvesh Kumar, Himanshu Gautam, Shivendra Singh, and Mohammad Shafeeq. Top vulnerabilities in cloud computing. *ECS Transactions*, 107(1): 16887, 2022.
30. Nnamdi Johnson Ogbuke, Yahaya Y. Yusuf, Kovvuri Dharma, and Burcu A. Mercangoz. Big data supply chain analytics: Ethical, privacy and security challenges posed to business, industries and society. *Production Planning & Control*, 33(2–3): 123–137, 2022.
31. Adrien Bécue, Isabel Praça, and João Gama. Artificial intelligence, cyber-threats and industry 4.0: Challenges and opportunities. *Artificial Intelligence Review*, 54(5): 3849–3886, 2021.

15 Dodging Security Attacks and Data Leakage Prevention for Cloud and IoT Environments

Ishu Gupta
National Sun Yat-Sen University

Ankit Tiwari
TechMatrix IT Consulting Pvt. Ltd.

Priya Agarwal
Amdocs Development Center India LLP

Sloni Mittal
Hewlett-Packard (HP)

Ashutosh Kumar Singh
National Institute of Technology Kurukshetra

CONTENTS

DOI: 10.1201/9781003321149-15

15.1 INTRODUCTION

Cloud computing acts as a backbone for the emerging techniques such as the Internet of Things, big data, cyber-physical system (CPS) in the computer science field Wei et al. (2016) where the interconnection links between workstations are formed either by means of wired network connections or remote connections Gupta et al. (2022). With its rapid advancement, there is a need for data security as well as maintainability over the network in the cloud and IoT environments Gupta et al. (2020). The confidential data which is being revealed to malicious users can outcome in a major setback in organizations as well as for an individual Zaghloul et al. (2019). The illegitimate users carry out the attack with the help of unlicensed tools in the system to endanger system privacy and integrity Zhang et al. (2019). Protecting the system from undesired hostilities, illegitimate activities, and loss is called Security Gupta et al. (2019). Data security within wireless networks is provided with the help of cryptographic techniques and tools. Security over data ensures confidentiality, authenticity, and integrity for the documents sent across the network which further raises the availability and classification issues for the documents that compromise the security to a further extent Gupta and Singh (2019). This can further lead to trade secrets, bank details, privacy of patients, health records, and security of accounts, and the list goes on due to inappropriate classification of documents in categories different from the one they belong which violates the confidentiality of the users. These breaches grow in accordance with their size along with action Gupta and Singh (2018). For instance, eBay, an eminent online shopping website experienced one of the major setback leaks in history when more than 145 million customers' personal details such as phone numbers, names, and email IDs were theft. As an outcome, a massive reset account password was carried out from the customer's end Singh and Gupta (2020).

To eliminate these problems executants and scholars have illuminated techniques and methodology to safeguard the confidential data which is mainly acknowledged by the term data leakage prevention systems (DLPSs) Kaur et al. (2017a). DLPSs mark the continuous investigation of confidential data which lacked standard security mechanisms such as intrusion detection and firewalls. The confidential data is endlessly supervised by DLPSs irrespective of data "in transit" or "in use". The subject matter and context of the private data are mainly analyzed by the DLPSs for such kind of detection. Studying components such as sender, receiver, format, time, and data size fall under context analysis Kaur et al. (2017b). The regular expression, fingerprinting, and statistical analysis of content fall under content analysis. Thus, the use of DLPSs is increasing for curative actions such as caution, chunks, encryptions, and audits for safeguarding these tasks.

For all confidential and nonconfidential data, comprehensive indexing is required. Additional storage is a major drawback, as well as processing competence, is needed. To search for acceptable and resemble documents, binary codes as a memory address (BCMA) are used which was capable of getting resembling documents within the insignificant time taken by locality sensitive hashing (LSH). The main constraint is that it is only capable of getting documents determined from specific topics like government borrowing and disaster. The perception is made when duplicate or contents of the intimate data is being retrieved, broadcasted, or preowned without the consent of the user. Identification takes place when duplicates or parts of the intimate data are monitored, retrieved, or transferred without validation. Statistical analysis, regular expressions, and fingerprinting are used to recognize duplicates of confidential data. The drawback of the regular expression is that it has a restricted extent so only rule-based items are in the ease of satisfaction Baker and McCallum (1998). Advanced fingerprinting is the fingerprints produced by the data hashing having some faults. To overcome the limitations of the advanced fingerprinting technique, a modified full-data fingerprinting was presented Shu et al. (2015). However, in the fingerprints, the major culpability is that they are unsafe and can be dodged with smaller changes to the authentic one: that is why to generate altered fingerprints, k-skip-n-gram came into existence. Even after the altercation, k-skip-n-gram is a vigorous way of detecting the authentic data (such as subtraction, addition, and synonyms). In this method, both intimidated and nonintimidated docs are managed to create fingerprints, in which intimidated docs create nonintimidated k-skip n-grams that will help in removing unwanted n-grams in the confidential documents. In almost all scenarios this suggested procedure outperformed the nominal fingerprinting method Alneyadi et al. (2016). Still, the major drawback of fingerprinting is the vulnerability of being different from the one suspected. This chapter deals with and scrutinizes the success achieved by the use of statistical analysis to identify intimate connotations of data. Term frequency-inverse document frequency (TF–IDF) is the weighting function for the term on which the DLP model is based. The proposed method is shown to be bewildered when compared to various stages.

The rest of the chapter is organized as follows: Section 15.2 entails the historical journey, state of artworks along with the research gaps identified and contributions made in the domain. Section 15.3 presents the proposed solution by explaining the operational flow along with its architecture. The details of the implementation environment achieved the outcome of the proposed scheme, and the comparisons with the existing techniques are described in Section 15.4 followed by the conclusions and future directions in Section 15.5.

15.2 LITERATURE SURVEY

15.2.1 HISTORICAL NOTES

The central feasible networking exposure of security was given by Marin consisting of traffic analysis, computer intrusion detection, and network monitoring aspects of network security. Information security has been categorized into three sections by Wu Kehe – network system security, data security, and network business security. Flauzac has represented a similar proposal for the execution of dispersed security solutions in a supervised collective way, often known by the name grid of security, ensured by section of device, which is reliable and inter-connections among workstations may be executed under system guidelines power Shabtai et al. (2012). It is also accepted as a speculative basis for enterprise auto-matic production systems as security defense. Wuzheng has established a scheme over security for mobile networks on the basis of public key infrastructure (PKI). Cryptographic tools and network security tools have also been established mode (counter with cipher block chaining message authentication code (CBC-MAC)), Advanced Encryption Standard (AES), and cipher-based message authentication code (CMAC) are the various feasible applications for the hurdle related to network security Chandra et al. (2014).

The crucial challenge for an organization is to transfer the message and docu-ments over the network in a confidential and protected manner. According to a survey, internal data leakage is the major reason for the threats to information security which consist of an estimated 29% confidential or responsive accidental data leakage, an estimated 16% theft of intellectual property, and an estimated 15% other thefts consisting of financial data, and customer information. Boley et al. (1999) estimated that internal threats cause more destruction than external threats, with an overall estimate of 67%. The other major challenge in an organization is the classification of data outcomes. An article entitled "Automatic Indexing: An Experimental Inquiry" was published which suggested a "technique for classifi-cation of documents automatically in accordance with their index/contents". The report was broadcasted by Gartner, in 2013 on "Enterprise content-aware DLP" in contrast to the report broadcasted by Forrester Wave in 2008. Many machine learning approaches like support vector machine (SVM) in which data belonging to a major amount is categorized. The high FPR was achieved by the Naive-based approach of data classifier Lewis (1998).

The security and data leakage threats are increasing due to the following reasons:

(a) Tremendous increase in the wireless area network and rapid growth for the in-motion device network usage. (b) Enterprise networking and security are being transformed by mobile and cloud devices. (c) Tremendous growth in the malware prac-tice is adversely affecting the security. (d) There is a vital role in active attacks such as modification, fabrication, and interruption of the messages. (e) The documents con-taining sensitive data are difficult to identify. (f) Data are distributed throughout the networks.

(g) The contribution is toward the digital data across the network. The significance and importance of security and leakage prevention are as follows: (a) it gradually increases the digital transactions over the network. (b) Confidentiality, authentic-ity, and integrity of the message are maintained. (c) Unauthorized users are unable to access the message sent across the network. (d) Sensitive data movement and usage can be monitored. (e) The exfiltration of data by the outsider hackers can be identified.

15.2.2 STATE OF ART

Network security is something that tends to become a new social circumstance in today's world. It can often lead to message leaks, message alteration, confidentiality, and authenticity issues over the network. The crime recently committed by Kevin Mitnick is one of the biggest crimes in the history of the world. Losses were source code from different companies and 80 million dollars in US intellectual property. This is the reason network security has taken a great spot in the world. To embrace the central idea and focus behind network security, Komal Gandhi has conducted a survey related to network security attacks Gandhi (2016). A slow internet connection can cause various attacks on networks. The attacks are classified mainly into two classes active attacks and passive attacks. Former are caused by manipulating the documents or information sent over the network. The attacks classified under active attacks are Denial-of-Service (DOS) attacks, Domain Name Service (DNS) spoofing, smurf attacks, etc. Passive attacks are the attacks in which the third party tries to gain access to confidential documents sent over the network. The attacks classified under the passive attacks are wiretapping port scanner and isle scan. Security protocols were not accomplished within the Transmission Control Protocol/ Internet Protocol (TCP/IP) communication profusion. Communication is required to make it more secure due to the expansion in the internet architecture Singh et al. (2011); Uchnár and Hurtuk (2017). Yan et al. performed an exploration of network security and stated some of the network security technologies based on authentication and data encryption technologies Yan et al. (2015). Some of the authentication technology is categorized as message authentication, identity authentication, etc. Data encryption technology in the paper included cryptography using symmetric and asymmetric keys individually. The former case makes use of a similar key for encrypting and decrypting the data at both sender's as well as receiver's sides. The problem identified in this symmetric key encryption is the secure transfer of the private key to both parties. While the latter used both public and private keys for securing the data. The message encrypted using a public key can be decrypted with the use of a private key conversely. Asymmetric key encryption is demonstrated as more secured and intricate. Unconventional feature linked to message authentication has been presented by Meena et al. in Meena and Jha (2016). The basic lookout of message authentication and requirements has been discussed. It is interpreted that data integrity and source authentication are the foremost presumptions of message authentication. Additionally, the different inferences and methodologies that exist for message authentication have been discussed. Afterward, the differences between public and private key-based message authentication have been depicted. Various problems related to lack of compatibility and key sharing are also described. Bellare et al. have presented two related schemes named hash-based message authentication code (HMAC) and nested construction-based message authentication code (NMAC). The security and efficiency are provisioned by using the cryptographic hash function Bellare et al. (1996). Their analysis consists of the consideration of generic Message Authentication Code (MAC) schemes. However, the studies show that if notable weaknesses are found in the MAC then the hash function needs to be relinquished

TABLE 15.1

Comparison of Network Security Outlooks

	Various Outlooks		
Metrics	Basics	Emphasis On	Definition
Security attack	Degrade information system	Influencing available data, replaying message, analyzing data traffic, etc.	An attack which endangers the security over information
Process for security	Formulated implementation to safeguard security	Data consistency, digital signature, access control etc.	Techniques to find, restore or avoid any attacks over security
Security service	Includes confidentiality, authenticity, integrity and other security parameters	Security endangered, processing data, and transmission of information	The services that upgrade the transfer process of information and data processing system. One or more security techniques can be employed

TABLE 15.2

Major Algorithms and Their Operations

	Algorithms		
Operations	Hashing Algorithm	Symmetric	Asymmetric
Electronic signature	SHA family	RC4	DSA
Encryption	–	DES	RSA
Key exchange	MD5	AES	DH

from significant as well as a huge range of other standards and known usage. The detailed examination is done on various cryptography techniques (security algorithms) such as hash function, symmetric and asymmetric technique, and different approaches used for detecting data semantics. Table 15.1 summarizes the comparison of network security outlooks, and Table 15.2 lists the state-of-the-art algorithms along with their operations that are used for securing the data.

15.2.3 RESEARCH GAPS AND CONTRIBUTIONS

The objective of this research work is to build a model which is capable of effectively and securely transmitting the message or information over the network and categorizing the transferred data for "in-use" or "in transit" state individually by overcoming the shortcomings of the major existing methods for securing the data that are demonstrated in Table 15.3 Semwal and Sharma (2017). The aim is to prevent the message from transmitting from one node to another in the communication channel without having any intervention from the third or unauthorized party in the network and safeguarding the confidentiality which might occur due to the wrong clustering of documents and leads

TABLE 15.3
Drawbacks of Existing Techniques

Operations	Technique	Limitations
Message encryption	Symmetric key	The key can be accessed by any intruder.
	Asymmetric key	Public key of sender which is used for decryption on receiver's site is vulnerable to access by another person on the network.
Message authentication code	Authentication without cryptography	No encryption technique performed over the message.
	Authentication tied to a plain text	Encryption of data is not conducted. The identical keys are used for both the purposes, enciphering and deciphering the data.
	Authentication tied to cipher text	Enciphering and deciphering the data is done using symmetric key.
Hash function	Authentication with plain text	Plain data is in unencrypted form.
	Authentication with encrypted text	Message is decrypted by employing public key and that is accessible to any intruder on network.

to revealing of the personal information to the outside source other than the intended user. The main goal of this work is to ensure the secure transmission of confidential messages or information over the network, without any alteration or modification, while maintaining confidentiality, authenticity, and integrity throughout the transmission and use. Only authorized parties should have access to the information, and the classification should align with predefined classes for the documents. Moreover, the documents that are newly generated in which confidential data might be contained can be recognized if they are classified as a confined category. Based on the semantic approach, a statistical data leakage prevention (DLP) model is presented to classify the data. It uses neighboring context and content of personal/private data to identify malicious access and secure the confidential data despite the data being used by the intended users or while data are transferred internally or externally between communication channels.

15.3 PROPOSED METHOD

The proposed solution allows the safe and secure transmission of a message from one network to another without the interruption from the third or unauthorized party where the known and authorized party only is permissible to access the message. The MITM (Man In The Middle) is incapable of performing the MITM attacks over the message as the method encrypts the messages twice and it becomes very difficult to have access to the keys or codes for the decryption.

15.3.1 Operational Flow

The proposed work achieves confidentiality, authenticity, integrity, reliability, and availability by applying the following techniques Kumari (2017); Michail et al. (2007).

15.3.1.1 Message Encryption

Using Symmetric Key: symmetric key encryption comes with the problem of key transmission. Secret keys have to be sent to the receiver workstation before transmitting the actual message. The channels being used for transmission are insecure as it is not possible to promise that the channel will not be accessed by anyone else. Therefore, exchanging the keys individually is the only secure and intact way. The entire functioning of encryption by employing a symmetric key is depicted in Figure 15.1.

Public Key Encryption Using Asymmetric Key: in this technique, data attains its privacy but loses its legitimacy. For instance, the receiving station B is supposed to get messages from station A only but all users in the network are having the public key of B so messages may be dropped to B from any other user in the network other than A. Therefore, receiving station B cannot be sure of the message being received from supposed user A or any other users in the network. Figure 15.2 presents the entire functioning of this encryption technique.

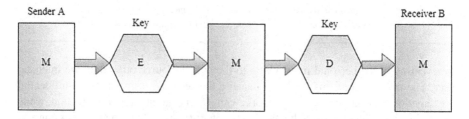

FIGURE 15.1 Encryption using symmetric key.

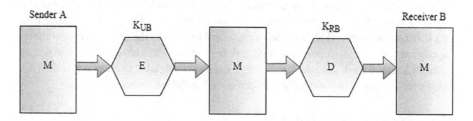

FIGURE 15.2 Asymmetric message encryption using public key.

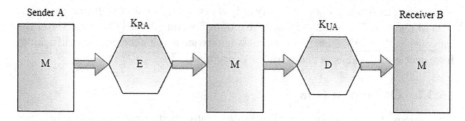

FIGURE 15.3 Asymmetric data encryption through private key.

Private Key Encryption Using Asymmetric Key: in this technique, legitimacy is attained but privacy is not achieved. Since the public key of the sender is known to everyone in the network therefore data transmitted by the sender station could be deciphered by employing its private key by everyone on the network other than the supposed receiver. Figure 15.3 represents the entire functioning of the asymmetric key encryption through private key Trnka et al. (2018).

15.3.1.2 Message Authentication Code

Here, no cryptography technique is applied to the data therefore confidentiality is unachievable. The data can be accessed by any illegitimate user on the network. Figure 15.4 represents the functioning of this scheme.

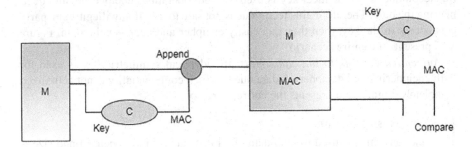

FIGURE 15.4 Process of message authentication code.

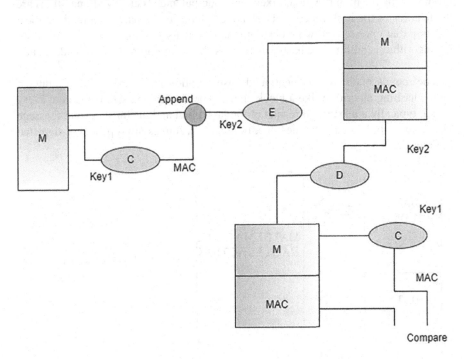

FIGURE 15.5 MAC without encryption.

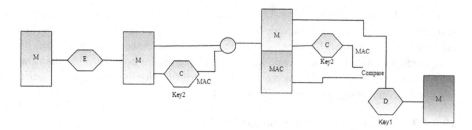

FIGURE 15.6 MAC with encryption.

Authentication link associated with plain data: the main issue associated with this technique is the identical key is used for both operations enciphering and deciphering the data. Therefore, this technique is not safe to use. If any illegitimate party gets access to the key, then they can easily encipher and decipher the data. Figure 15.5 presents the entire scenario.

Authentication link to the cipher text: the identical symmetric key is used for both enciphering and deciphering data due to which confidentiality is not actualized completely. Figure 15.6 presents the entire scenario.

15.3.1.3 Hash Function

The function which is used to map data of whimsical size into a determined size is called a hash function. Hash values, digests, and hash codes are the values that are given back by the hash function. Fixed length output and efficiency of operations are the features of the hash function. Hash code is built on the basis of a mathematical function that operates on two blocks of data of fixed size. The portion of the hashing algorithm is built by the hash function, as shown in Figure 15.7 Alkandari et al. (2013).

The rounds of the above-mentioned hash function in Figure 15.7 are indulged by the hashing algorithm like a block cipher. The determined size is taken at every round, typically a mixture of results of the last round and the most current message block. To hash the complete message, the required rounds are repeated as depicted in Figure 15.8.

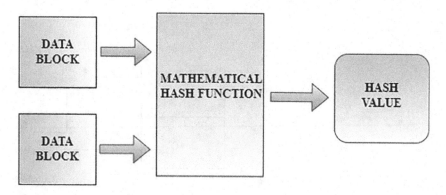

FIGURE 15.7 Block diagram of hash function.

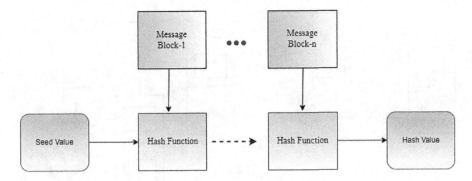

FIGURE 15.8   Schematic representation of hashing algorithm.

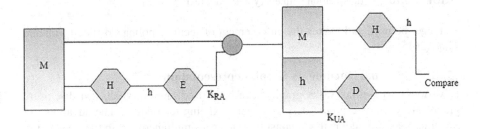

FIGURE 15.9   Authentication without encrypted plain text.

Authentication of hash code without ciphered plain data: the plain data are not in encrypted form; therefore, confidentiality is endangered and when this plain data is adjoined with hash code as shown in Figure 15.9, it is vulnerable to access by an intruder on the network. Also, the public key used for the decryption of data is accessible to everyone on the network, therefore, authenticity is also endangered and we cannot ensure the decryption of data by only the supposed receivers or anyone else.

Authentication with encrypted plain text: as decryption is done by employing a public key, authenticity is not realized as depicted in Figure 15.10. The sent information is accessible to everyone on the network and it is difficult to ensure if it is the supposed recipient or anyone else.

15.3.1.4   Textual Documents Clustering

The clustering textual documents problem is related to high dimensional clustering. Every document is assumed to be a vector containing 1,000 terms and every term of the textual document in an individual dimension. In order to classify each and every document related to resemblance angle, a cosine measure between two documents is calculated. Support vector machine (SVM) that uses the SVM algorithm is used to categorize three types of data: enterprise private, enterprise public, and nonenterprise. Detection of data leakage is 97% with 3% false negative. But this method only identifies private and public categories and failed to detect more flexible classification levels like top secret, secret, and confidential. Thus, this method can

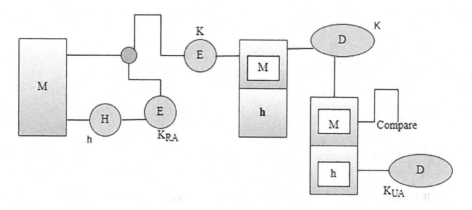

FIGURE 15.10   Authentication with encrypted plain text.

no longer be used and makes the enforcement of security policies difficult Van and Thuc (2015).

15.3.1.5   Textual Documents Graph Representation

In various text similar, tasks graphs have been used, due to which text disclosure graphs are used as a model despite various existing techniques. The advantage of the graph-based model is that it constitutes a term simultaneous with their context in addition to monitoring the context along with document structure.

15.3.1.6   Depletion of Redundant Information

Various methods for the depletion of duplicate information exist such as latent semantic indexing (LSI), principal component analysis (PCA), and singular value decomposition (SVD). The task operating efficiency is mainly reduced by the duplicate information which is incompetent and the worst case is in text-related tasks.

15.3.2   Data Leakage Prevention Model

Data leakage is defined as a strategy that guarantees sensitive and confidential data will not go outside the organization. To detect confidential data semantics, statistical data analysis is used for DLP. The most-popular term weighting function named TF–IDF is employed to measure the tested document and the amount of information contained in it. The predefined categories cluster the related topic together which is basically considered to be the main aim of clustering. Considering that every category has a confidential level, documents with restricting secrecy levels can be easily detected, and accordingly, actions like blocks, alerts, and quarantines are taken. The document D can be easily identified with the use of fingerprinting and regular expression if a user U is attempting malicious access, use, and alter D. However, the user U alter the document D by inserting, deleting, or exchanging words, lines, or paragraph. Even in the case of robust data fingerprinting, identifying altered document

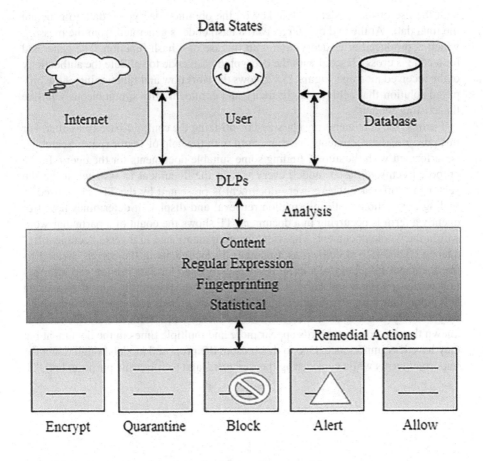

FIGURE 15.11   Deployment of data leakage avoidance model.

D is a challenging event. Therefore, statistical analysis is employed to calculate and approximate the private data with the help of the DLP model. The deployment model showcases the various data states, analyzing the various DLP context and various remedial actions, which have been taken is represented in diagrammatic form in Figure 15.11 for a better understanding of the prevention model.

15.3.3 ARCHITECTURE OF SECURE DATA TRANSMISSION

In the proposed solution, plain data are first encrypted by employing the sender's private key to generate the ciphertext. This ciphertext produces hash code from the hash function. Besides the copy of generated ciphertext that is adjoined together with the hash code is being used and it is encrypted again with the public key of the receiver for the stronger security of the document. On the receiver side, the private key of the receiver is used to decrypt the ciphertext, and then the decryption is applied again with the use of the sender's public key on the obtained decrypted data to generate the final data. At the end of the receiver, a hash code is generated from the message which is produced after decryption with the use of a hash function. The generated hash code is cross-checked with the decrypted hash code to validate the authenticity of the received message. Figure 15.12 shows the workflow and functioning of the proposed solution that achieves authenticity and confidentiality simultaneously unlike the existing schemes.

Further, the documents are classified by utilizing the centroid-based classifier. For this purpose, the documents are showcased with the help of vector space. Searching is performed with the aim of finding some suitable documents for the query. In the proposed centroid-based model, every text in the document is assumed to be the vector in term space where every document is presented by the TF. TF is used in finding connections with information retrieval and display. It determines how frequently a term is occurring in a document. TF shows the count of a particular word within the whole document. This value is often mentioned in inverse document frequency (IDF). IDF is defined as a number of documents in the corpus divided by the document frequency of a term. It is basically used for data mining and information retrieval Cohen and Hirsh (1998). IDF plays an important role in measuring how important a term is. Since TF considers all the terms equally important, that's why TF only is not sufficient to calculate the weight of a term in a document. It is known that the common words appear more and multiple times in the document but they have less importance. Due to this reason, the proposed models weigh down the important terms while considering the rare ones, and to solve this purpose, both TF

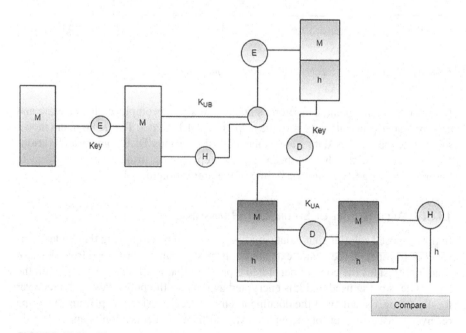

FIGURE 15.12 Proposed architecture.

and IDF are combined together and named TF–IDF in the proposed solution to get an ultimate score of term t in a document d. Further, the stop list is utilized to remove similar words.

15.4 PERFORMANCE EVALUATION

15.4.1 EXPERIMENTAL SET-UP

The implementations are performed on a machine assembled with two Intel® Xeon® Silver 4114 CPU@2.20 GHz 40 core processor, 64-bit Ubuntu 16.04 LTS, and 128 GB RAM with the following required installation: i Text Editor ii Java Development Kit 17.0.2. The implementation scenario is considered for the data security performance evaluation where the two users named Parle and Britania have interchanged their public keys. A confidential file named "conf.txt" needs to be sent from Parle to Britania. The comprehensive features of numerous documents are gathered for the experiments of DLP. For the performance evaluation of the DLP phase, the data sets Badminton (Badm), Soccer (Socc), Cricket (Crick), and Tennis (Tenn) are used in the experimental results. The considered data sets are collected from Badminton (2022); Cricket (2022); Soccer (2022); Tennis (2022), respectively. To prepare data for analysis, they are first converted into text format. The categorization of data from various algorithms on these data sets is performed. Approximately, 80% documents are randomly selected as the training set while the remaining 20% documents are taken as a test set. The average precision outcome from 10 runs has been presented in this chapter. k-Nearest neighbor (KNN), C4.5, Naive Bayesian (NB), and proposed centroid-based categorization have been performed on the same platform. Furthermore, the centroid-based proposed solution interpretation has been assessed and validated by contrasting it against numerous numbers of document-gathering techniques named KNN, NB, and C4.5.

15.4.2 DATA SECURITY OUTCOME AND COMPARISON

Figure 15.13 shows the sample of the confidential message which needs to be protected during transmission in a cloud environment. For the protection of this message, encryption is performed. Public and private keys are generated for Parle, as shown in Figures 15.14 and 15.15, respectively. Similarly, the public and private keys for Britania are also generated as depicted in Figures 15.16 and 15.17, respectively.

The following are the steps performed for encryption: (i) a new symmetric or secret key is generated as displayed in Figure 15.18. (ii) The file "conf.txt" is encrypted with the help of the obtained key in the first step. (iii) The key obtained in step 1 is encrypted

```
1 Hey! How are you Britania?
```

FIGURE 15.13   Sample of confidential message.

FIGURE 15.14 Parle private key.

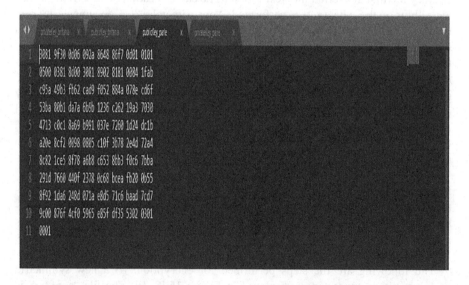

FIGURE 15.15 Parle public key.

by using Britania public key for the stronger protection where the encrypted secret key is portrayed in Figure 15.19. (iv) Parle transfers both the attained encrypted key and encrypted message to Britania that are portrayed in Figures 15.19 and 15.20, respectively.

Britania obtains two files and performs the decryption over these files to generate the original message. The following steps are performed by Britania for decryption:

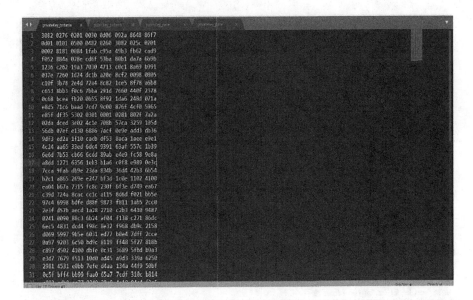

FIGURE 15.16 Britania private key.

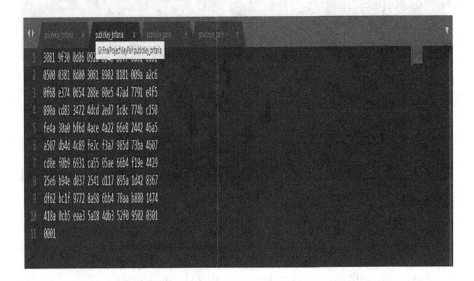

FIGURE 15.17 Britania public key.

1. It performs the decryption with the help of its own private key over the obtained encrypted secret key and generates the decrypted secret key as shown in Figure 15.21.
2. The attained encrypted message is decrypted using the secret key obtained in step 1 and the plain text is obtained as depicted in Figure 15.22.

FIGURE 15.18 Secret key.

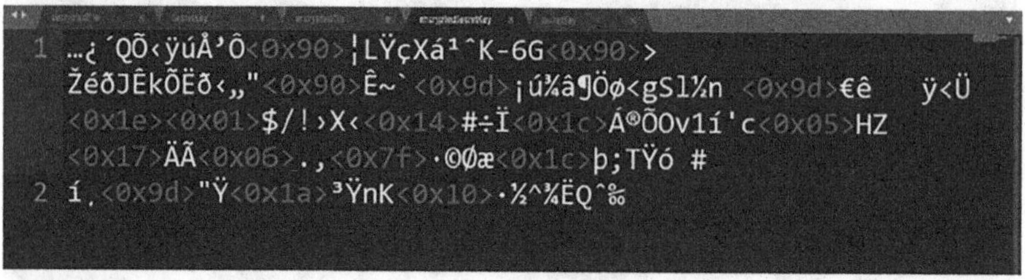

FIGURE 15.19 Encrypted secret key.

FIGURE 15.20 Encrypted data.

FIGURE 15.21 Decrypted secret key.

Furthermore, Table 15.4 lists the comparison of the proposed scheme with the state-of-the-art techniques. It can be seen that no other models except the proposed solution provide both authentication and confidentiality simultaneously.

15.4.3 DLP Outcome and Comparisons

To evaluate the DLP outcome, the data classification is performed over the four datasets named badminton Badminton (2022), cricket Cricket (2022), soccer Soccer (2022), and tennis Tennis (2022) and compared with the existing KNN and NB approaches. The original and predicted outcomes obtained from the existing classifiers named KNN, Naive Bayesian, and centroid-based proposed solution are reflected in Figures

FIGURE 15.22 Generated text.

TABLE 15.4

Comparison of Proposed and Other Considered Methods

Technique	Key	Mode	Authentication	Confidentiality
Message encryption	Symmetric key	–	Yes	No
	Asymmetric key	Public key encryption	No	Yes
		Private key encryption	Yes	No
Message authentication code	Symmetric key	Authentication link to plain text	Yes	No
		Authentication link to ciphertext	Yes	No
Hash function	Asymmetric key	Hash function without encrypted text	No	No
		Hash function with encrypted text	No	Yes
Proposed solution	Asymmetric key	–	Yes	Yes

15.23 and 15.24 and Table 15.5, respectively where all the three schemes are implemented at the same platform. It can be seen that the centroid-based proposed scheme categorizes the data correctly unlike the existing classifiers because of maintaining the resemblance between the specific class and a test document. Furthermore, these outcomes are used to compute the accuracy for measuring the performance of the model, which is presented in Table 15.6.

The categorization of data from various algorithms on multiple data sets is performed. Table 15.6 depicts the accuracy of data categorization achieved by the proposed scheme and comparison with the existing classifiers KNN and NB. It can be seen that the proposed scheme which is based on a centroid

classifier outperforms both the Naive Bayes and KNN for all the data sets, i.e., Cricket (Crick), Tennis (Tenn), and Soccer (Socc) except Badminton (Badm). Centroid-based classifier outperforms various algorithms and performs well in all the cases due to regulating the resemblance between the specific class and a test document.

Differentiation between various classification algorithms using some sample pairs tested values is listed in Table 15.6. Here, the analytical notable results are

```
In [35]:  from sklearn.cluster import KMeans
          kmeans = KMeans(n_clusters=4, random_state=0).fit(X)
          labels_ = kmeans.labels_
```

```
In [36]:  print(len(b),len(c),len(s),len(t))
          labels_
```

          26 28 24 12

```
Out[36]:  array([3, 3, 3, 3, 3, 3, 3, 3, 3, 3, 3, 3, 3, 3, 3, 3, 3, 3, 3, 3, 3, 3,
                 3, 3, 3, 3, 0, 0, 0, 0, 0, 0, 0, 0, 0, 0, 0, 0, 0, 0, 0, 0, 0, 0,
                 0, 0, 0, 0, 0, 0, 0, 0, 0, 0, 2, 2, 2, 2, 2, 2, 2, 2, 2, 2, 2, 2,
                 2, 2, 2, 2, 2, 2, 2, 2, 2, 2, 2, 2, 1, 1, 2, 1, 1, 1, 1, 1, 1, 1,
                 1, 1])
```

```
In [37]:  Y_pred = kmeans.predict(Y)
          Y_pred
```

```
Out[37]:  array([3, 3, 3, 3, 3, 3, 3, 3, 3, 3, 3, 3, 3, 3, 3, 3, 3, 3, 3, 3, 3, 3,
                 3, 3, 3, 3, 0, 0, 0, 0, 0, 0, 0, 0, 0])
```

FIGURE 15.23   Outcome of k-means algorithm.

```
Out[38]:  MultinomialNB(alpha=1.0, class_prior=None, fit_prior=True)
```

```
In [39]:  nb.predict(test_new)
```

```
Out[39]:  array([0, 0, 0, 0, 0, 0, 0, 0, 0, 1, 1, 1, 1, 1, 1, 1, 1, 1, 2, 2, 2, 2, 2,
                 2, 2, 2, 2, 2, 2, 3, 3, 3, 3, 3, 3, 3, 0], dtype=int64)
```

```
In [40]:  Ytest_actual = encoder.transform(test_label)
          print(Ytest_actual)
```

          [0 0 0 0 0 0 0 0 0 1 1 1 1 1 1 1 1 1 2 2 2 2 2 2 2 2 2 3 3 3 3 3 3 3 3]

FIGURE 15.24   Outcome of Naive Bayesian algorithm.

summarized using the sample paired test in which various classification algorithms are taken. It compares the performance of the two classifiers. It is shown whether a classifier is performing best or worse than another classifier. It shows that a better performance is given by the row classifier than the classifier represented in the column. It is found that the centroid-based proposed scheme is three times better than NB and worse in one data set. Alike, the centroid-based proposed scheme is three times better than C4.5 and one time worse in all four data sets. Overall, the centroid classifier-based proposed scheme is unpredictably good than the other existing

TABLE 15.5
Comparison of Actual and Predicted Labels

Sr. No.	Original	Predicted	Sr. No.	Original	Predicted
0	Badminton	Badminton	18	Soccer	Soccer
1	Badminton	Badminton	19	Soccer	Soccer
2	Badminton	Badminton	20	Soccer	Soccer
3	Badminton	Badminton	21	Soccer	Soccer
4	Badminton	Badminton	22	Soccer	Soccer
5	Badminton	Badminton	23	Soccer	Soccer
6	Badminton	Badminton	24	Soccer	Soccer
7	Badminton	Badminton	25	Soccer	Soccer
8	Badminton	Badminton	26	Soccer	Soccer
9	Cricket data	Cricket data	27	Soccer	Soccer
10	Cricket data	Cricket data	28	Tennis	Tennis
11	Cricket data	Cricket data	29	Tennis	Tennis
12	Cricket data	Cricket data	30	Tennis	Tennis
13	Cricket data	Cricket data	31	Tennis	Tennis
14	Cricket data	Cricket data	32	Tennis	Tennis
15	Cricket data	Cricket data	33	Tennis	Tennis
16	Cricket data	Cricket data	34	Tennis	Tennis
17	Soccer	Soccer	35	Tennis	Tennis

TABLE 15.6
Comparative Analysis of the Proposed and Other Considered Methods

Algorithms	Categories			
	Cricket	Tennis	Badminton	Soccer
Naive Bayesian	89.3	91.2	84.3	72.3
KNN	85.8	87.5	77.5	84.6
Proposed scheme	91.8	93.9	82.7	94.2

classifiers such as KNN, NB, and C4.5 since it regulates the resemblance between the specific class and a test document. The mean resemblance is calculated between the test document and all the additional documents in the current class. If the magnification is higher, it corresponds to a small level of resemblance between the documents, whereas if the magnification is lower, it corresponds to a high level of resemblance among the documents.

15.5 CONCLUSIONS AND FUTURE WORK

The malicious affairs are rising expeditiously during transmission on the network, therefore increased security is required on the network for the data being transferred.

For this, authenticity and confidentiality must be maintained to forward the messages from sender to receiver in the most consistent and safe form. This chapter has reviewed some of the common existing techniques, highlighted the limitations, and provided an integrated solution to overcome the existing problems. The proposed solution achieved both confidentiality and authenticity simultaneously unlike the existing scheme with the use of asymmetric key and applying double encryption and decryption to secure the message from intruder's undesirable access. Now, the data can be securely transferred; however, still, data leakage can take place. Data leakage can happen because of using bad clustering approaches. To prevent data leakage, an approach based on a centroid document classifier is proposed in which the data is clustered appropriately which helps in DLP. Centroid-based document classification is steady and maintainable, and it outperforms other classification algorithms on various datasets that prove the validity of the proposed model.

In the future, the emerging quantum machine learning algorithms could be incorporated with the work for improved accuracy in identifying and preventing data leakage incidents. Furthermore, the malicious entity could be predicted in advance using a quantum computing mechanism to strongly protect the data against data leakage occurrences.

ACKNOWLEDGMENT

This work is supported by the University Grants Commission (UGC), Ministry of Human Resource Development (MHRD), and Government of India under the scheme of National Eligibility Test-Junior Research Fellowship (NET-JRF) with grant no. F.15-9(JUNE 2015)/2015(NET).

REFERENCES

Alkandari, A. A., Al-Shaikhli, I. F., & Alahmad, M. A. (2013, September). Cryptographic hash function: A high level view. In *2013 International Conference on Informatics and Creative Multimedia* (pp. 128–134). IEEE.
Alneyadi, S., Sithirasenan, E., & Muthukkumarasamy, V. (2016). A survey on data leakage prevention systems. *Journal of Network and Computer Applications*, 62, 137–152.
Badminton (2022). Badminton Dataset. https:badminton-information.com
Baker, L. D., & McCallum, A. K. (1998, August). Distributional clustering of words for text classification. In *Proceedings of the 21st Annual International ACM SIGIR Conference on Research and Development in Information Retrieval* (pp. 96–103). ACM.
Bellare, M., Canetti, R., & Krawczyk, H. (1996, August). Keying hash functions for message authentication. In *Annual International Cryptology Conference* (pp. 1–15). Springer.
Boley, D., Gini, M., Gross, R., Han, E. H. S., Hastings, K., Karypis, G., ... & Moore, J. (1999). Document categorization and query generation on the world wide web using webace. *Artificial Intelligence Review*, 13(5), 365–391.
Chandra, S., Paira, S., Alam, S. S., & Sanyal, G. (2014, November). A comparative survey of symmetric and asymmetric key cryptography. In *2014 International Conference on Electronics, Communication and Computational Engineering (ICECCE)* (pp. 83–93). IEEE.
Cohen, W. W., & Hirsh, H. (1998, August). Joins that generalize: Text classification using WHIRL. In *KDD* (pp. 169–173). AAAI Press.

Cricket (2022). Cricket Dataset. https:cricbuzz.com

Gandhi, K. (2016, March). Network security problems and security attacks. In *2016 3rd International Conference on Computing for Sustainable Global Development (INDIACom)* (pp. 3855–3857). IEEE.

Gupta, I., & Singh, A. K. (2018). A probabilistic approach for guilty agent detection using bigraph after distribution of sample data. *Procedia Computer Science*, 125, 662–668.

Gupta, I., & Singh, A. K. (2019). Dynamic threshold based information leaker identification scheme. *Information Processing Letters*, 147, 69–73.

Gupta, I., Singh, N., & Singh, A. K. (2019). Layer-based privacy and security architecture for cloud data sharing. *Journal of Communications Software and Systems*, 15(2), 173–185.

Gupta, I., Gupta, R., Singh, A. K., & Buyya, R. (2020). MLPAM: A machine learning and probabilistic analysis based model for preserving security and privacy in cloud environment. *IEEE Systems Journal*, 15(3), 4248–4259.

Gupta, I., Singh, A. K., Lee, C. N., & Buyya, R. (2022). Secure data storage and sharing techniques for data protection in cloud environments: A systematic review, analysis, and future directions. *IEEE Access*, 10, 71247–71277.

Kaur, K., Gupta, I., & Singh, A. K. (2017a, August). A comparative evaluation of data leakage/loss prevention systems (DLPS). In *Proc. 4th Int. Conf. Computer Science & Information Technology (CS & IT-CSCP)* (pp. 87–95).

Kaur, K., Gupta, I., & Singh, A. K. (2017b, December). Data leakage prevention: e-mail protection via gateway. In *Journal of Physics: Conference Series* (Vol. 933, No. 1, p. 012013). IOP Publishing.

Kumari, S. (2017). A research paper on cryptography encryption and compression techniques. *International Journal of Engineering and Computer Science*, 6(4), 20915–20919.

Lewis, D. D. (1998, April). Naive (Bayes) at forty: The independence assumption in information retrieval. In *European Conference on Machine Learning* (pp. 4–15). Springer.

Meena, U., & Jha, M. K. (2016, March). A retrospective investigation of message authentication in wireless sensor networks: A review. In *2016 3rd International Conference on Computing for Sustainable Global Development (INDIACom)* (pp. 613–616). IEEE.

Michail, H. E., Kakarountas, A. P., Selimis, G., & Goutis, C. E. (2007, July). Throughput optimization of the cipher message authentication code. In *2007 15th International Conference on Digital Signal Processing* (pp. 495–498). IEEE.

Semwal, P., & Sharma, M. K. (2017, September). Comparative study of different cryptographic algorithms for data security in cloud computing. In *2017 3rd International Conference on Advances in Computing, Communication & Automation (ICACCA)(Fall)* (pp. 1–7). IEEE.

Shabtai, A., Elovici, Y., & Rokach, L. (2012). *A survey of data leakage detection and prevention solutions*. Springer Science & Business Media.

Shu, X., Yao, D., & Bertino, E. (2015). Privacy-preserving detection of sensitive data exposure. *IEEE Transactions on Information Forensics and Security*, 10(5), 1092–1103.

Singh, A. K., & Gupta, I. (2020). Online information leaker identification scheme for secure data sharing. *Multimedia Tools and Applications*, 79(41), 31165–31182.

Singh, S. K., Singh, M. P., & Singh, D. K. (2011). A survey on network security and attack defense mechanism for wireless sensor networks. *International Journal of Computer Trends and Technology*, 1(2), 9–17.

Soccer (2022). Soccer Dataset. https:eurosoccerfan.com

Tennis (2022). Tennis Dataset. tennis-ontheline.com

Trnka, M., Cerny, T., & Stickney, N. (2018). Survey of authentication and authorization for the internet of things. *Security and Communication Networks*, 2018, 1–18.

Uchnár, M., & Hurtuk, J. (2017, January). Safe user authentication in a network environment. In *2017 IEEE 15th International Symposium on Applied Machine Intelligence and Informatics (SAMI)* (pp. 000451–000454). IEEE.

Van, D. H., & Thuc, N. D. (2015, August). A privacy preserving message authentication code. In *2015 5th International Conference on IT Convergence and Security (ICITCS)* (pp. 1–4). IEEE.

Wei, J., Liu, W., & Hu, X. (2016). Secure data sharing in cloud computing using revocable- storage identity-based encryption. *IEEE Transactions on Cloud Computing*, 6(4), 1136–1148.

Yan, F., Jian-Wen, Y., & Lin, C. (2015, June). Computer network security and technology research. In *2015 Seventh International Conference on Measuring Technology and Mechatronics Automation* (pp. 293–296). IEEE.

Zaghloul, E., Zhou, K., & Ren, J. (2019). P-mod: Secure privilege-based multilevel organizational data-sharing in cloud computing. *IEEE Transactions on Big Data*, 6(4), 804–815.

Zhang, L., Cui, Y., & Mu, Y. (2019). Improving security and privacy attribute based data sharing in cloud computing. *IEEE Systems Journal*, 14(1), 387–397.

16 Role of Blockchain in Industry 4.0

Keshav Kaushik and Bhavana Kaushik
University of Petroleum & Energy Studies

CONTENTS

16.1 INTRODUCTION TO BLOCKCHAIN TECHNOLOGY: CHARACTERISTICS AND PROTOCOLS

A blockchain is a distributed, decentralized database of records that allows for quick, secure transactions without being monitored by a centralized administration. The future of cryptocurrencies is anyone's guess, but it is important to recognize the importance of blockchain technology in a number of financial and nonfinancial areas. Blockchain is a development of linked data structures, or blocks, that store or trace everything that happens in distributed environments in a distributed organization. Every block is joined to the preceding block by a unique pointer known as a hash pointer, culminating in a chain and a structure made up of annexes: a permanent and unchangeable history that may be used as a reviewing trail by any member to continuously verify the accuracy of the records by simply analyzing the data itself.

On a blockchain (Zheng et al., 2017), a digital signature that attests to its legitimacy protects every transaction. The information saved on the blockchain is tamper proof and cannot be altered thanks to the use of encryption and digital signatures. Consensus, which is the process of all network members coming to an accord, is made possible by blockchain technology. Every piece of information kept on a blockchain is digitally recorded and has a shared history that is accessible to everyone on the network. By doing this, any possibility of fraudulent or transaction repetition is avoided without the use of a third party.

Imagine a situation in which you are trying to find a way to transmit some funds to a buddy who lives somewhere in order to better grasp blockchain. A bank or a money transfer service like PayPal or Paytm are two options you can often employ. With this choice, third parties are needed to complete the transaction, which results in a cost that is added to your money as a transfer fee. Additionally, in situations like these, it is impossible to guarantee the security of your funds because it is very likely that

DOI: 10.1201/9781003321149-16

a hacker may interrupt the network and take your money. The victim in both situations is the client. Blockchain is useful in this situation. If we utilize a blockchain in these situations rather than a bank to transfer money, the procedure is significantly simpler and more reliable. As the money is processed directly by you, there is no additional cost as a third party is not required. Additionally (Gamage et al., 2020), the blockchain system is decentralized and not restricted to a particular place, making all the data and records maintained to be public and decentralized. No danger of data tampering by a hacker exists since the data is not kept in a single location. There are several characteristics of blockchain that makes it a promising technology. The blockchain's characteristics are shown in Figure 16.1.

- **Immutable:** any data that have been added to a blockchain cannot be modified after it has been done so, which is known as the immutability feature of a blockchain. Attempt sending an email as an illustration to better grasp immutability. An email that has been sent to a large group of recipients

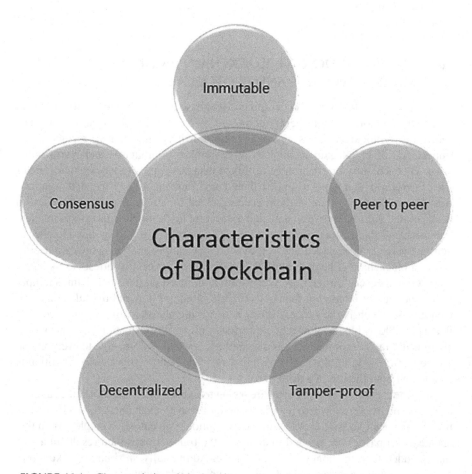

FIGURE 16.1   Characteristics of blockchain.

cannot be canceled. You will have to ask each receiver of your email to erase it, which is a laborious workaround. This is the operation of immutability. Data cannot be modified or amended once it has been processed. Because each block in a blockchain retains the hash of the one before it, if you try to modify the data of one block, you will have to update the whole blockchain that follows it. Any changes to one hash will affect the remaining hashes as well. Since it takes a lot of processing power to modify all the hashes, it is quite difficult for somebody to do so. Data stored in a blockchain is unaffected by modifications or hacker attacks due to irreversibility.

- **Peer to Peer:** the usage of blockchain makes it simple to engage between two parties using a peer-to-peer architecture without the need for a middleman. Blockchain (Kaushik et al., 2020) is a peer-to-peer protocol that enables each member of the network to have an exact copy of each transaction, allowing machine consensus for authorization. For instance, with blockchain, you may complete any transactions from one region of the world to the other in a matter of seconds. Any additional costs or delays won't be deducted from the transaction either.

- **Tamper Proof:** because blockchains have the immutability characteristic built in, it is simpler to spot data manipulation. Because any alteration to even a single block can be easily discovered and corrected, blockchains are thought to be tamper proof. Hashes and blocks are the two main methods for spotting tampering. Each block-related hash function is distinct, as was previously mentioned. It is comparable to the fingerprint of a block. Any alteration to the information will cause the hashing algorithm to change as well. A hacker would have to modify the hashes of all the blocks following that one in addition to making any modifications because the hash function of one block is connected to the next, which is a challenging task.

- **Decentralized:** Since blockchains are decentralized, no one individual or organization has control over the whole network. The distributed ledger is accessible to everybody on the internet, but nobody is able to independently edit it. This particular aspect of blockchain technology provides safety and transparency while offering people control.

- **Consensus:** Each blockchain has an agreement to assist the network in reaching choices quickly and impartially. Consensus (Dhar Dwivedi et al., 2021) is a decision-making technique that helps the network's active nodes swiftly come to a consensus and ensures the system runs smoothly. Although nodes might not have much confidence in one another, they might have confidence in the network's central algorithm. There are numerous accessible consensus techniques, each having advantages and disadvantages. A consensus method is necessary for any blockchain or else it will start to lose value.

It is crucial to realize that there are hundreds of procedures in use, making it impossible to research all of the available possibilities in a reasonable length of time. The most crucial protocols, nevertheless, are the following four: a summary of the most common protocols used in blockchain development solutions is provided below.

- **Multichain:** in order to promote more effective transactions and to generate new implementations for the proof-of-work methods that blockchain networks depend on, multichain was founded to assist for-profit businesses in building private blockchains. As a privately held business, multichain is able to provide an Application Programming Interface (API) that blockchain development services may utilize to simplify collaboration and hasten implementation. The way that multichain is built to function with fiat currencies and actual repositories of value sets it distinct from its rivals. On the other hand, the majority of cryptocurrency initiatives are focused on the eventual replacement of traditional currency with digital forms of trade.

- **Corda:** Corda is a rival of multichain that provides a protocol tailored for businesses. The financial and banking industries have seen the most Corda application development. Nevertheless, Corda's technology may be applied to a wide range of unique blockchain applications. Corda is a suitable option for blockchain development alternatives in the finance sector because it is certified by the R3 banking consortium.

- **Hyperledger:** An open-source initiative called Hyperledger (Kaushik et al., 2022) intends to develop a set of tools that businesses may use to install blockchain technology rapidly and efficiently. The protocol's libraries, which speed up implementation, make it a popular choice for blockchain software applications. Strongly supporting Hyperledger, the Linux Foundation has contributed considerable expertise to hasten the development of the standard. Due to the fact that Linux and Hyperledger are extremely interoperable, these servers are often employed in the modern corporate environment.

- **Ethereum:** A variant of Ethereum's software geared for commercial use cases is available. The objective of Ethereum Enterprise is to expand the commercial use cases for blockchain technologies. Businesses may quickly create extensive apps to trade value with Ethereum Enterprise. The main benefit of Ethereum Enterprise is that it enables companies to develop their own private variations of Ethereum while still utilizing the most recent Ethereum technology. Ethereum's license makes it challenging to create proprietary versions of the program under normal conditions, but the commercial edition offers enterprises a workaround.

16.2   CHALLENGES AND SOLUTIONS OF BLOCKCHAIN TECHNOLOGY IN INDUSTRY 4.0

One of the upcoming mobile technologies that Industry 4.0 will develop with is blockchain. Blockchain technology has the potential to improve security, confidentiality, and data transparency both for small and large organizations. Modern manufacturing practices that accelerate corporate goals are combined to form Industry 4.0. Research has recently focused on the potential impact of a number of Industry 4.0 technologies, including artificial intelligence (AI), IoT, big data, and blockchain.

These technologies give the industrial and supply chain sectors a wide choice of possibilities. Blockchain is a digital ledger that has attracted a lot of attention and can improve the environment for manufacturing and supply chains. The benefits of blockchain now are fascinatingly understood in a variety of areas. The significant potential of blockchain in Industry 4.0 is discussed in this article (Javaid et al., 2021). For perspectives, a number of blockchain technology's drivers, facilitators, and related capabilities are examined.

Blockchain is a tamper-proof solution with several uses. The secret of blockchain (BC) is its tamper-proof design, which gives data authentication reliability. The principles of blockchain and how it functions in numerous different applications and technology are the main topics of this paper (Kaushik & Dahiya, 2022). The readership will also learn more about problems with blockchain technology integration in this study. The paper will outline the role that blockchain technology plays in solving problems related to cutting-edge technologies. This analysis finds peer-reviewed material that attempts to leverage blockchain for cybersecurity goals and provides a rigorous study of the most often used blockchain security features. IoT, security, wireless communications, 5G, and even beyond networks are just a few of the industries where the revolutionary blockchain has a wide range of applications. This article discusses how blockchain technology will be used in 5G and other future networks. The many difficulties that arise when using the blockchain in 5G and future networks are also highlighted in this article (Kaushik, 2022). The report also covers how blockchain will be used in 5G networks in the future.

To increase the levels of security of the smart applications, numerous security benchmarks and remedies have been suggested over the years; however, the current solutions are either based on centralized architectures with single points of failure or have high communication and computation costs. Additionally, the majority of the security solutions now available have a narrow focus and do not address scalability, resilience, storage systems, network latency, traceability, data integrity, and provenance. Blockchain technology may provide a remedy for the aforementioned problems. These facts served as the impetus for the rigorous assessment of various blockchain-based technologies and their suitability for various Industry 4.0-based applications that we offer in this article (Bodkhe et al., 2020).

The fourth industrial revolution, also referred as Industry 4.0, is on the horizon, and as a result, new disruptive innovations are being taken into account for integration in the manufacturing setting. A few of these options is the blockchain, which attempts to handle business transactions, connect disparate systems, and promote asset traceability. As a result, this innovation helps to build an efficient supply chain that might have an influence on the worldwide market. The intersection between blockchain and Industry 4.0 is revealed in this article (Silva et al., 2020). In order to provide instances of what is now being suggested, promoted, and produced, scientific, corporate, and governmental efforts are investigated in this manner.

Innovative upcoming technologies like the 5G mobile network and blockchain can fill these demands. In contrast to blockchain's novel data-sharing method of operation, which improves the high degree of security, visibility, and trustworthiness of stored data, 5G will enable extraordinarily large channel capacity and lower data latency. As a result (Jovović et al., 2019), this study provides a broad overview of the prospective use cases for the 5G network and blockchain technology in Industry 4.0. The findings might help early adopter businesses acquire market technical dominance.

The growth (Lypnytskyi, 2019) of blockchain's decentralized versions is most consistent with the idea of Industry 4.0, according to research on the technology's progress. Nevertheless, as demonstrated in the article, internal conflicts caused by the blockchain "trilemma" and ineffective energy utilization make it difficult to apply such variants in the industrial setting. There are ways to break the "blockchain trilemma" deadlock, along with some cutting-edge, contemporary alternatives like the Level-2 standard. Blockchain technology (BCT) can significantly contribute to industrial sustainability by safeguarding resources, protecting the environment, and improving citizen quality of life. In this paper, a four-class taxonomy of the most significant cyberattacks in Industry 4.0 during the past ten years is offered. Industry 4.0's most significant BCT-based works are compared in terms of their secrecy, integrity, scalability, security, and multifactor identification capabilities. Our analysis demonstrates that BCT integration in the industry can guarantee integrity and confidentiality and should be mandated to maintain data availability and confidentiality. Prospective (ElMamy et al., 2020) research topics are outlined with a focus on implementing BCT in the industrial setting while taking into account machine learning, 5G systems, and other evolving technologies.

Distribution and decentralization are seen as common architectural elements. Aoun et al. (2021) aim to broaden people's perspectives on blockchain technology's potential as a liberating instrument for the fourth industrial revolution. We are first examining the foundations of Industry 4.0, its difficulties, constraints, and opportunities, and then we are looking at ways that blockchain technology might offer new capabilities and worth to the implementation of Industry 4.0.

16.3 FUTURE RESEARCH SCOPE OF BLOCKCHAIN IN INDUSTRY 4.0

Blockchain technology plays a crucial role in the Industry 4.0 and has its wide range of applications. This section of the chapter highlights the future scope of blockchain in industry 4.0 by comparing the major findings of the latest research work done in this domain. Table 16.1 shows the major findings of scope of blockchain in industry 4.0.

16.4 CONCLUSION

The book chapter highlights the protocols and characteristics of blockchain technology. The chapter will also enlighten the readers about the challenges and solutions of blockchain technology in Industry 4.0. Moreover, the chapter also discusses the future research scope of blockchain technology in Industry 4.0. The chapter will be helpful for blockchain enthusiasts, blockchain developers, financial market experts, Ph.D. scholars, researchers, and students.

TABLE 16.1

Major Findings of Scope of Blockchain in Industry 4.0

Authors	Year	Major Findings
Alladi et al. (2019)	2019	In addition to major commercial blockchain networks in various pertinent industries, this report offered the most recent research findings in each of the connected industrial sectors. The writers talked about the difficulties that each sector will face in using blockchain.
Swami et al. (2021)	2021	The authors of this study made an effort to analyze some of the most recent IoT applications in a variety of industries, including sports, smart gadgets, location monitoring, and blockchain-based security. IoT is still a subject that is fast expanding and has a lot of room for multidisciplinary study and research. IoT will continue to grow as sensors are integrated into more systems.
Hassan Onik et al. (2018)	2018	Both a blockchain-based human resource management system algorithm and a recruitment system management have been presented. It is clear from the examination of the case study data that the suggested methodology has distinct benefits over the current recruitment methods.
Zuo (2020)	2020	In-depth information about blockchain systems, structures, methodologies, and research problems is presented in this study. Our talks on utilizing blockchain technology for different implementations of the smart factory and smart supply chain are guided by the authors' suggested blockchain architecture for smart manufacturing.
Leng et al. (2021)	2021	This research shows how the literature has investigated various cybersecurity-related challenges. Future research paths for blockchain-secured smart factory are offered based on the insights gleaned from this analysis, which may serve as a roadmap for research on pressing cybersecurity issues for reaching intelligence in Industry 4.0.

REFERENCES

Alladi, T., Chamola, V., Parizi, R. M., & Choo, K. K. R. (2019). Blockchain applications for Industry 4.0 and industrial IoT: A review. *IEEE Access, 7*, 176935–176951. https://doi.org/10.1109/ACCESS.2019.2956748

Aoun, A., Ilinca, A., Ghandour, M., & Ibrahim, H. (2021). A review of Industry 4.0 characteristics and challenges, with potential improvements using blockchain technology. *Computers & Industrial Engineering, 162*, 107746. https://doi.org/10.1016/J.CIE.2021.107746

Bodkhe, U., Tanwar, S., Parekh, K., Khanpara, P., Tyagi, S., Kumar, N., & Alazab, M. (2020). Blockchain for Industry 4.0: A comprehensive review. *IEEE Access, 8*, 79764–79800. https://doi.org/10.1109/ACCESS.2020.2988579

DharDwivedi, A., Singh, R., Kaushik, K., Rao Mukkamala, R., & Alnumay, W. S. (2021). Blockchain and artificial intelligence for 5G-enabled Internet of Things: Challenges, opportunities, and solutions. *Transactions on Emerging Telecommunications Technologies*, e4329. https://doi.org/10.1002/ETT.4329

ElMamy, S. B., Mrabet, H., Gharbi, H., Jemai, A., & Trentesaux, D. (2020). A survey on the usage of blockchain technology for cyber-threats in the context of Industry 4.0. *Sustainability 2020, 12*(21), 9179. https://doi.org/10.3390/SU12219179

Gamage, H. T. M., Weerasinghe, H. D., & Dias, N. G. J. (2020). A survey on blockchain technology concepts, applications, and issues. *SN Computer Science 2020, 1*(2), 1–15. https://doi.org/10.1007/S42979-020-00123-0

HassanOnik, M. M., Miraz, M. H., & Kim, C. S. (2018). A recruitment and human resource management technique using blockchain technology for industry 4.0. *IET Conference Publications, 2018*(CP747). https://doi.org/10.1049/CP.2018.1371

Javaid, M., Haleem, A., Pratap Singh, R., Khan, S., & Suman, R. (2021). Blockchain technology applications for Industry 4.0: A literature-based review. *Blockchain: Research and Applications, 2*(4), 100027. https://doi.org/10.1016/J.BCRA.2021.100027

Jovović, I., Husnjak, S., Forenbacher, I., & Maček, S. (2019). Innovative application of 5G and blockchain technology in Industry 4.0. *EAI Endorsed Transactions on Industrial Networks and Intelligent Systems, 6*(18), e4–e4. https://doi.org/10.4108/EAI.28-3-2019.157122

Kaushik, K. (2022). Demystifying blockchain in 5G and beyond technologies. *Journal of Mobile Multimedia, 18*(5), 1379–1398. https://doi.org/10.13052/JMM1550-4646.18513

Kaushik, K., & Dahiya, S. (2022). *Scope and Challenges of Blockchain Technology*. Springer Nature, 461–473. https://doi.org/10.1007/978-981-16-8248-3_38

Kaushik, K., Dahiya, S., & Sharma, R. (2022). Role of blockchain technology in digital forensics. *Blockchain Technology*, 235–246. https://doi.org/10.1201/9781003138082-14

Kaushik, K., Dahiya, S., Singh, R., & Dwivedi, A. D. (2020). Role of blockchain in forestalling pandemics. *Proceedings -2020 IEEE 17th International Conference on Mobile Ad Hoc and Smart Systems, MASS 2020*, 32–37. https://doi.org/10.1109/MASS50613.2020.00014

Leng, J., Ye, S., Zhou, M., Zhao, J. L., Liu, Q., Guo, W., Cao, W., & Fu, L. (2021). Blockchain-secured smart manufacturing in Industry 4.0: A survey. *IEEE Transactions on Systems, Man, and Cybernetics: Systems, 51*(1), 237–252. https://doi.org/10.1109/TSMC.2020.3040789

Lypnytskyi, D. V. (2019). Opportunities and challenges of blockchain in industry 4.0. *Economy of Industry, 1*(85), 82–100. https://doi.org/10.15407/ECONINDUSTRY2019.01.082

daSilva, T. B., de Morais, E. S., de Almeida, L. F. F., da Rosa Righi, R., & Alberti, A. M. (2020). *Blockchain and Industry 4.0: Overview, Convergence, and Analysis*. Springer, 27–58. https://doi.org/10.1007/978-981-15-1137-0_2

Swami, M., Verma, D., & Vishwakarma, V. P. (2021). Blockchain and industrial Internet of Things: Applications for Industry 4.0. *Advances in Intelligent Systems and Computing, 1164*, 279–290. https://doi.org/10.1007/978-981-15-4992-2_27/COVER/

Zheng, Z., Xie, S., Dai, H., Chen, X., & Wang, H. (2017). An overview of blockchain technology: Architecture, consensus, and future trends. *Proceedings - 2017 IEEE 6th International Congress on Big Data, BigData Congress 2017*, 557–564. https://doi.org/10.1109/BIGDATACONGRESS.2017.85

Zuo, Y. (2020). Making smart manufacturing smarter – A survey on blockchain technology in Industry 4.0. *Enterprise Information Systems, 15*(10), 1323–1353. https://doi.org/10.1080/17517575.2020.1856425

17 Blockchain and Bitcoin Security in Industry 4.0

Yadunath Pathak and Praveen Pawar
Visvesvaraya National Institute of Technology

CONTENTS

17.1 INTRODUCTION

Blockchain, which Satoshi Nakamoto created to serve as Bitcoin's public transaction log, has gained much attention [1]. Blockchain technology allows for decentralized transaction processing. It is currently employed in a number of industries, including financial services, Internet of things (IoT) applications, reputation management, and others. The security and scalability of blockchain are just two of the problems that still need to be resolved.

Cryptocurrency is the buzzword in business and academia these days. Bitcoin, a very popular cryptocurrency, has a market value of over $10 billion [2]. Bitcoin has a distinctive data storage structure, and transactions are carried out alone. Bitcoin's core technology is blockchain. In 2008, the concept of blockchain was developed, and in 2009, it was realized [1]. Blocks are used to store all transactions on the blockchain. It is referred to as the public ledger. With the addition of new blocks, the chain

DOI: 10.1201/9781003321149-17

grows longer. A distributed consensus mechanism and asymmetric cryptography algorithms are used to keep the ledger consistent and ensure user security. Blockchain technology has important aspects of decentralization, auditability, consistency, and anonymity. By combining all of these features, blockchain reduces costs and increases efficiency.

Blockchain applications may be utilized in various financial services such as remittance, digital assets, and online payment because blockchain payments can be made without a middleman [3]. It can be utilized in a variety of domains, including security services [4], public services, smart contracts [5], IoT [6], and reputation systems [7]. These fields favor the blockchain in a variety of ways. Ciaian et al. [8] attempted to make it worldwide money by evaluating Bitcoin's properties. Kleineberg and Helbing [9] demonstrated how digital variety could be maintained. In a blockchain, transaction information cannot tamper with the original data. Businesses that require a high level of privacy and security using blockchain to attract customers. It employs a distributed computing model to avoid the occurrence of a single point of failure. When a smart contract is activated on the blockchain, it is carried out automatically. Regarding the internet platform, the blockchain has a slew of technological Zchallenges. Bitcoin is not suited for high-frequency trading because its network can only handle seven transactions per second. As a result, the larger the block, the more storage space is required, resulting in slower network diffusion. Hence, decentralization in the blockchain is required. The trade-off between security and size is the most challenging task. Second, miners can increase their profits using a selfish mining technique [10]. Miners disguise their blocks to increase their profits.

Blockchain development hinders because it allows for frequent branching. If users utilize a private key [11] and a public key, privacy leakage might occur in blockchain. In addition, the current proof-of-stake or proof-of-work consensus algorithm has several flaws. For example, proof-of-work wastes much electrical energy, whereas proof-of-stake consensus allows you to get wealthy faster. Wikis, journal articles, blogs, forum postings, and codes are all examples of blockchain literature. A technological analysis was conducted for the decentralized digital currencies known as Bitcoin [12].

Real-time information is required in the Industry 4.0 environment to create a seamless service and production system. Processing time is the limiting component, hence practical applications must be properly taken into account. Blockchain is the ideal technology to keep records and information because it can overcome several obstacles. Therefore, the purpose of this study is to determine how blockchain might fit into Industry 4.0. In order to improve transaction efficiency in the company, blockchain replenishment is now necessary. Industry 4.0's sustainable product lifecycle management is made possible by blockchain technology. This will thrive and give processes a higher level of safety in production environments with lower risks. Before organizations adopt blockchain for implementation, though, there is still a lot of work to be done with the platform. Although the risk can be reduced, organization executives must recognize the positive effects of this technology on their business because it is always developing. Industry 4.0 involves a higher level of privacy and faith. The next section presented the architecture of blockchain.

17.2   BLOCKCHAIN ARCHITECTURE

In contrast to a single central server, a blockchain is a network of interconnected computers. Computers in a network must agree with the state of their shared data and abide by its restrictions. Every blockchain's block has some effect on the shared state. A blockchain is a set of interconnected blocks that serves as a public ledger and records all transactions. Blockchain's blocks have only one parent block, and the block header contains the hash of the preceding block. The Ethereum blockchain is made up of uncle blocks, or the offspring of a block's ancestors. The first block in a blockchain without any parents is known as the genesis block [13]. As shown in Figure 17.1, a block comprises a block body and a block header. The block header is made up of the following elements:

- Block Version: it specifies the block validation processes to be followed.
- Merkle Tree Root Hash: this is made up of the hash values for all transactions in a block.
- Timestamp: this is a number that represents the current time in seconds.
- n Bits: this is the minimum size of a valid hash block.
- Nonce: this field is four bytes long. It starts at zero and grows in value with each calculation.
- Parent Block Hash: with a 256-bit hash value, this field points to the previous block.

Transactions and the transaction counter are found in the block's body. The size of each transaction and the size of the block determine the number of transactions in a block. Asymmetric cryptography is used to validate the transaction authenticity [14]. Asymmetric cryptography-based digital signatures are employed in unreliable

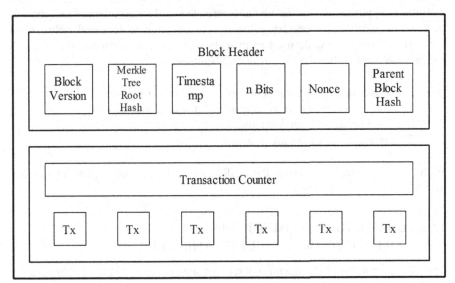

FIGURE 17.1   Block diagram of a blockchain.

environments. It is important to know how Bitcoin works, which is explained in the next section.

17.3 HOW BITCOIN WORKS?

Bitcoin is built on the peer-to-peer transaction concept, which means that the owner only has access to a public address and a private key. It indicates that the transaction is peer to peer and does not involve a financial institution, and that the timestamp of transactions is hashed using the network. The hashes are then connected to create the blockchain, which is a chain of hashes. The Bitcoin network, which is its foundation, acts as a permanent record or ledger of all transactions observed there. Cryptographic data are generated using the SHA-256 method. The private key of the Bitcoin's owner is used to sign each transaction. Thus, unauthorized interference is avoided.

Two items are required for Bitcoin transactions: (i) A Bitcoin address and (ii) A private key that only the owner knows about it. An asymmetric key pair is formed by these two things. The public key that makes up the Bitcoin address is made up of a random string of numbers and letters. User A's private key is used to sign the input (transaction source and amount) before using user B's public key to sign the output, which is then transmitted to user B as a private message. The Bitcoins are then transferred to user B's wallet. To complete a sequence of independent verification, it takes roughly 10 minutes. The Bitcoin miners will provide these confirmations. The network's shared consensus structure provides confidence to Bitcoin transactions through the mining process. Miners are at the heart of the Bitcoin system, and they're in charge of ensuring that all transactions on the network are genuine. A block is added to the blockchain for each new Bitcoin production using the hash function.

Bitcoin cryptocurrency is the first decentralized, and independent of any centralized control (such as a monetary authority). Users perceive decentralization when the verification of the code used to provide services is transparent. However, its decentralized nature has the following disadvantages:

* Every user maintains a public ledger.
* There is no central authority to certify distributed transactions.
* Bitcoin production is anonymous.
* Bitcoin exchange values are dynamic.

With all these noticeable points, the challenges in the Bitcoin implementation are discussed in the next section.

17.4 CHALLENGES IN THE BITCOIN
IMPLEMENTATION: SECURITY THREATS

The working concept of Bitcoin must first be grasped to better understand the security threats to Bitcoin. As a result, it has become the focus of modern study and the topic of in-depth studies [15]. Government legislation and the power of traditional currencies are two more nontechnical elements that influence Bitcoin's value. This chapter aims to figure out which of the causes described above will produce significant variations in the Bitcoin currency value.

In 2017, Fraser and Bouridane compiled a list of potential security concerns about Bitcoin [16] and an analysis of the factors that contributed to the price drop. The following are the details:

17.4.1 BITCOIN PROTOCOL

There are always risks as Bitcoin is purely an online commodity. More than 50% of the Bitcoin network can be taken over by a single person or group of users, thanks to a security flaw that Bradbury nicknamed the "51 percent attack" [17]. According to Bonneau et al. [18], the problem of the size of the Bitcoin network (millions of nodes) is extremely challenging to handle. Even if it is not a major threat, Bitcoin is nonetheless subject to it. Moreover, the value of Bitcoin is impacted by public apprehension.

17.4.2 BITCOIN SERVICES

The number of services using Bitcoin increased in tandem with its popularity. B¨ohme et al. [19] suggested the two public sectors that hold and trade Bitcoins: digital wallet services and currency exchanges. A digital wallet, according to their definition, is a computerized representation of the accounts, transactions, and private keys needed to send or use Bitcoins. In addition, rather than going through the time-consuming process of mining, people can buy Bitcoin through a Bitcoin exchange [20]. The Bitcoin system is less susceptible to cyberattacks of any kind than the aforementioned Bitcoin services.

17.4.3 OTHER FACTORS

Other factors that influence the value of Bitcoin include financial decisions and government regulations, in addition to security breaches. Because Bitcoin is a form of currency, any issues with traditional currencies will also apply to Bitcoin. It is noted that the most influential element was security threats to the Bitcoin protocol and their research was regarded to be the most effective method for determining the most important element influencing Bitcoin's value.

17.5 RISE OF BITCOIN VALUE

Despite the fact that Bitcoin's value fluctuates frequently, it remains the most valuable cryptocurrency. Bitcoin had a fantastic year in 2019, with more individuals starting to trust the technology and spend money on it. By the middle of 2019, its worth had risen to $20,000. Only 21 million Bitcoins are available for use around the world. With 4 million still to be mined, just 17 million Bitcoins have already been created, further increasing the value of the cryptocurrency. Bitcoin reached a peak in 2021 of roughly $47,300. Bitcoin and other cryptocurrencies have piqued the curiosity of countries such as the United States, Japan, and South Korea.

We are aware that blockchain is a decentralized network, that offers security, anonymity, transparency, and integrity. Despite the absence of a central authority to confirm and validate the transactions, every blockchain transaction is regarded as being 100% safe and validated. The consensus protocol, a crucial component of any blockchain network, is the fundamental mechanism that makes this feasible. In the next section, consensus algorithms are discussed in detail.

17.6 CONSENSUS ALGORITHMS IN BLOCKCHAIN

The Byzantine Generals Problem transition is brought up in relation to the blockchain. A party of command generals and some of the Byzantine army encircled the city in the event of the Byzantine Generals Problem [21]. There were two points of view among the command generals. Some wanted to flee, while others wanted to assault. It would, however, collapse if only a few of the generals attacked. As a result, agreement on whether to attack or withdraw must be established. Similarly, obtaining a consensus in a distributed setting is difficult. As a result, because blockchain is a distributed environment, it faces the same difficulty. In blockchain, there is no central node that determines the similarity of ledgers on dispersed nodes. Some protocols must be followed to ensure the consistency of ledgers among nodes [22]. This section goes through the many strategies for reaching consensus in blockchain. Proof-based consensus algorithms are described in this section.

17.6.1 PROOF OF WORK (PoW)

The Bitcoin network employs PoW as a consensus method [1]. In a decentralized network, someone must be chosen to keep track of transactions. Although the random selection method is the easiest, it is also the most vulnerable. Therefore, a time-consuming procedure is needed to validate a node for publishing transactions in order to make sure that it won't attack the network. The block header's hash value is determined by each node. Nonce and miners comprise the head of a block. The nonce is frequently changed, giving the miners a variety of hash values. That consensus insures the resulting value must be less than or equal to the provided value. One node will reach the goal value once. A block is broadcast to the other nodes to validate the hash value. When a new block is verified, the miners upload it to their own blockchain. The nodes that calculate the hash values are known as miners, and Its PoW method is referred to as mining in Bitcoin.

When numerous nodes and the right nonce are chosen at almost the same moment, a valid block is formed in a decentralized network. As seen in Figure 17.2, branches can be generated in a variety of ways.

It is, however, impossible for two forks to develop branches at the same time. The chain is legitimate in the PoW protocol if it becomes long after that. The forks generated by blocks V4 and A4 that were both validated at the same time were considered. Miners execute the mining until a longer chain is discovered. A4, A5, and A6 make up the lengthier chain. As a result, the V4 miners move to the longer chain. In PoW, miners conduct large computations, yet they squander several resources. Specific PoW protocols with work-related side applications have been developed to

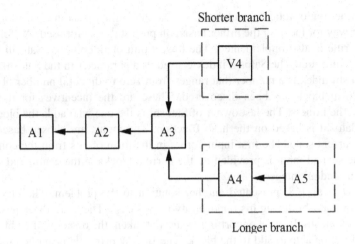

FIGURE 17.2   Two scenarios of blockchain branches.

compensate for the loss. Primecoin [22], for example, is used in mathematical studies to find a particular chain of prime integers.

According to the analysis, one of PoW's drawbacks is that there may be some security or usage difficulties. As a result, many remedies to these limits are proposed. The speed at which the blocks are added to the puzzles is increasing daily, raising the difficulty of the riddles even more. As a result, miners must spend on the gear to be the first to solve the mystery. Miners with lesser investment levels are unable to meet market demands. Tromp [23] presented a way for replacing puzzle labor with the Cuckoo Hash Function, which requires less effort from miners and allows for more accessible block adding.

17.6.2 Proof of Stake (PoS)

PoS is a low-energy alternative to PoW. In a PoS system, miners must show ownership of the currency amount. The people who have a huge quantity of currencies are less inclined to assault the network. Choosing a candidate based on account balance is unfair because the richest individual dominates the network. As a result, numerous ways combining stake size to determine the next block to be forged have been presented. BlackCoin [24], in particular, uses randomization to forecast the next generation. The formula takes the lowest hash value and multiplies it by the stake size. Peercoin prefers a coin-based selection mechanism. In Peercoin [25], the older and larger set of coins have a better chance of mining the next block. In comparison to PoW, PoS saves energy and is more efficient. Because mining is free, an attack could occur as a result. In many blockchains, PoW is used at first, and afterward, PoS is implemented. Ethereum, for example, is planning to switch from Ethash (a PoW type) to Casper (a PoS type) [26].

17.6.3 Pure Stake-Based Consensus

PoS in its purest form is NextCoin [27]. The bigger the stake a miner has, the more chances he or she has of mining a block. Consider, "a" number of coins available to the miner out of total "n" number of coins, then miner has a/n chance of mining. Another way for locating the miner based on pure stake is proposed. A "follow-the-satoshi" rule is also implemented. The lowest unit of Bitcoin, the Satoshi value, is taken into account. The Satoshi index is used as a parameter in the Satoshi method. The Satoshi index is a number that ranges from zero to the total number of Satoshi. The Satoshi block has been discovered. These are the incentives for the block's creators, the miners. The last owner of Satoshi is the one who adds the block to the chain. Satoshi is based on the hash function. The inputs are chosen based on the chain's current state. The first input comes in the form of bits from the comb function. The second input is provided by the current blocks in the chain, and the third input is a random integer.

Bentov et al. [28] presented another solution to the problem. The block is not formed by Satoshi during his opportunity in this case. There are three possibilities to choose Satoshi. If these opportunities are not taken, the owner will be blacklisted and will not be able to add to the block. The Bentov principle is applied to the PoS consensus [29] through the follow-the-satoshi approach. It specifies that the leader is selected at random using the entropy value. This determined value must be safeguarded because it would be difficult to replicate the protocol and forecast the value in order to influence the leader's election. On the basis of the Bentov election [30], the entropy value is determined. Epoch is taken, which is a picture of each stakeholder. Each epoch determines the amount of stake holders who will participate in the coin-flipping procedure.

17.6.4 PROOF OF ELAPSED TIME (PoET)

On permissioned blockchain networks, PoET is one of the best consensus methods [31]. The permissioned network makes decisions based on concepts of voting or mining rights. The network as a whole is transparent. Miners are identified before joining, so the system login is safe. As a result, the winners are chosen only on the basis of fairness. In this case, an individual's waiting time restriction is set at random. A ledger can be assigned to the player who completes their fair share of waiting time in order to construct a new block. The Intel SGX system [32] is an example of this.

17.6.5 PROOF OF ACTIVITY (PoA)

PoA is a hybrid of PoW and PoS algorithms [33]. It has the advantages of being more secure against attacks and not being a power-hungry system. Miners solve a critical puzzle for a payout, similar to the PoW algorithm. Miners are just interested in the block templates. A block's header specifies a set of random stakeholders. Stakeholders with higher stakes have a better chance of getting blocks approved.

17.6.6 PROOF OF BURN (PoB)

To secure cryptocurrency, a percentage of the coins are burned in the PoB process. The eater's address receives some coins. The coins supplied to the eater's address are nonreturnable. Burned coins are recorded in the ledger, making them really unspendable. Although there is a loss when coins are burned, the damage is just temporary because the currencies are protected against hackers in the long run. It also raises the stakes for other coins. This raises the likelihood of mining the upcoming blocks and the payments [34].

17.7 VARIOUS SOCIAL BLOCKCHAIN APPLICATIONS

- Business: traditional systems are more error-prone and move more slowly. In order to settle disputes that are stressful, expensive, and time-consuming, intermediaries are required. Asset Management: Trade Processing and Settlement [35] is an example of a business. For cross-border transactions, the old method is dangerous and costly. Records are kept inefficiently and with inaccuracies. The encryption of records in blockchain ledgers reduces these inaccuracies. Insurance claims are another example. These are more prone to errors because the processors must manually review bogus claims and process forms. Blockchain allows for risk-free administration and transparency. Money laundering is possible in the worldwide payment sector, which is also prone to errors. Blockchain startups such as Align Commerce, Bitspark, and Abra are providing solutions to this challenge. The blockchain has been combined with a payment application [36].
- Smart Property: any tangible or intangible asset, including real estate titles, stock certificates, patents, homes, and automobiles, can be equipped with smart technology. The registration information, as well as the terms of the property ownership contract, can be kept in a ledger. Smart property reduces the danger of fraud while also improving efficiency and trust.
- Blockchain IoT: the IoT refers to a network of interconnected objects. The Internet of things has a variety of effects on the user. When a printer's cartridges run out, for example, they can be ordered instantly from any shopping website using the printer. Smart appliances, supply chain sensors, and other smart devices can all benefit from blockchain and IoT [37].
- Smart Contracts: it includes autonomous digital machines with built-in "if this, then that" (IFTTT) coding. The intermediate makes sure that protocols are followed in real life. Blockchain eliminates the use of an intermediary by making all contract information available to users. It has applications in financial derivatives, insurance premiums, and other related applications. Additionally, it might be applied to blockchain-based healthcare. Using a private key, all individual medical records may be kept and safeguarded on the blockchain. Only those with specific access privileges would have access. Drug monitoring, healthcare supply management, and test results can all be tracked using the ledger [38].
- Blockchain Identity: by encrypting data and safeguarding it from scammers, the blockchain secures the user's identity. A passport is one of the examples. In 2014, the digital passport was introduced to assist users in identifying themselves. The blockchain ensures the accuracy of birth, death, and anniversary data. It encrypts personal information such as birth and death dates[39].

- Financial Services: the use of blockchain technology, also known as distributed ledger technology, to share data in a safe and effective manner may help the banking sector improve business procedures Distributed Ledger Technology (DLT). As a major mechanism for digital monetary transactions, blockchain technologies are altering the financial services sector. Profit pools and company models that are now inefficient will be exposed to the possibilities of inefficient blockchain financial platforms. Using blockchain technology, the distributed ledger increases the security and transparency of back-office activities. From a regulatory and auditing standpoint, this is preferred. The current capital market infrastructure is both costly and slow. Intermediates are required. The new blockchain innovations for capital markets are streamlining processes and lowering costs.
- Government: government services can be also improved with the applications of blockchain. It promotes the openness of government-citizen relations. It improves company processes by allowing for safe data sharing. It has a wide range of applications in the public sector. Blockchain technology can be used by governments to enhance service delivery, cut costs, do away with red tape, and stop tax fraud. Health care, social security benefits, and other fields are significantly impacted by the usage of blockchain in government processes. The processes used by the government are streamlined.

17.8   CONCLUSION

One of the most important aspects of Industry 4.0 is   digitization, which allows businesses to benefit from effectiveness in all areas, from technology and management consulting to supply business model and strategies. According to the findings, security vulnerabilities in Bitcoin services have the greatest impact on Bitcoin's value. Bitcoin's benefits have exceeded its drawbacks. As a result, Bitcoin has gained acceptance throughout the world, and many countries are now interested in its technology. According to a Nasdaq estimate, the stock's predicted value by the end of the year will be $23,499. Bitcoin is off to a strong start. People would be able to use their blockchain identities for a variety of tasks, ranging from basic actions to apps, software, and digital signatures. Blockchain may be the alternative to facilitating this step by providing a reliable source for high-quality interactional experience and understanding to smaller businesses. The volatility of Bitcoin value will be reduced if security breaches consequently weaken the Bitcoin services, hence the protocols need to be carefully designed.

REFERENCES

1. Satoshi Nakamoto. Bitcoin: A peer-to-peer electronic cash system. *Decentralized Business Review*, page 21260, 2008.
2. Garrick Hileman. State of blockchain q1 2016: Blockchain funding overtakes bitcoin. *CoinDesk*, New York, NY, May, 11, 2016.

3. George Foroglou and Anna-Lali Tsilidou. Further applications of the blockchain. In *12th Student Conference on Managerial Science and Technology*, volume 9, 2015.

4. Benjamin W. Akins, Jennifer L. Chapman, and Jason M. Gordon. A whole new world: Income tax considerations of the bitcoin economy. *Pittsburgh Tax Review*, 12:25, 2014.

5. Ahmed Kosba, Andrew Miller, Elaine Shi, Zikai Wen, and Charalampos Papamanthou. Hawk: The blockchain model of cryptography and privacy-preserving smart contracts. In *2016 IEEE Symposium on Security and Privacy (SP)*, pages 839–858. IEEE, 2016.

6. Yu Zhang and Jiangtao Wen. An IoT electric business model based on the protocol of bitcoin. In *2015 18th International Conference on Intelligence in Next Generation Networks*, pages 184–191. IEEE, 2015.

7. Mike Sharples and John Domingue. The blockchain and kudos: A distributed system for educational record, reputation and reward. In *European Conference on Technology Enhanced Learning*, pages 490–496. Springer, 2016.

8. Pavel Ciaian, Miroslava Rajcaniova, and d'Artis Kancs. The economics of bitcoin price formation. *Applied Economics*, 48(19): 1799–1815, 2016.

9. Kaj-Kolja Kleineberg and Dirk Helbing. A "social bitcoin" could sustain a democratic digital world. *The European Physical Journal Special Topics*, 225(17): 3231–3241, 2016.

10. Ittay Eyal and Emin Gun Sirer. Majority is not enough: Bitcoin mining is vulnerable. In *International Conference on Financial Cryptography and Data Security*, pages 436–454. Springer, 2014.

11. Alex Biryukov, Dmitry Khovratovich, and Ivan Pustogarov. Deanonymisation of clients in bitcoin p2p network. In *Proceedings of the 2014 ACM SIGSAC Conference on Computer and Communications Security*, pages 15–29, 2014.

12. Florian Tschorsch and Björn Scheuermann. Bitcoin and beyond: A technical survey on decentralized digital currencies. *IEEE Communications Surveys & Tutorials*, 18(3): 2084–2123, 2016.

13. Vitalik Buterin et al. A next-generation smart contract and decentralized application platform. *White Paper*, 3(37): 2-1, 2014.

14. Zibin Zheng, Shaoan Xie, Hong-Ning Dai, Xiangping Chen, and Huaimin Wang. Blockchain challenges and opportunities: A survey. *International Journal of Web and Grid Services*, 14(4): 352–375, 2018.

15. Feroz Ahmad Ahmad, Prashant Kumar, Gulshan Shrivastava, and Med Salim Bouhlel. Bitcoin: Digital decentralized cryptocurrency. In *Handbook of Research on Network Forensics and Analysis Techniques*, pages 395–415. IGI Global, 2018.

16. John Gregor Fraser and Ahmed Bouridane. Have the security flaws surrounding bitcoin effected the currency's value? In *2017 Seventh International Conference on Emerging Security Technologies (EST)*, pages 50–55. IEEE, 2017.

17. Danny Bradbury. The problem with bitcoin. *Computer Fraud & Security*, 2013(11): 5–8, 2013.

18. Joseph Bonneau, Andrew Miller, Jeremy Clark, Arvind Narayanan, Joshua A. Kroll, and Edward W. Felten. Sok: Research perspectives and challenges for bitcoin and cryptocurrencies. In *2015 IEEE Symposium on Security and Privacy*, pages 104–121. IEEE, 2015.

19. Rainer Böhme, Nicolas Christin, Benjamin Edelman, and Tyler Moore. Bitcoin: Economics, technology, and governance. *Journal of Economic Perspectives*, 29(2):213–238, 2015.

20. Tyler Moore and Nicolas Christin. Beware the middleman: Empirical analysis of bitcoin-exchange risk. In *International Conference on Financial Cryptography and Data Security*, pages 25–33. Springer, 2013.

21. Leslie Lamport, Robert Shostak, and Marshall Pease. The byzantine generals problem. In *Concurrency: The Works of Leslie Lamport*, pages 203–226. 2019.

22. Christopher Burks. Bitcoin: Breaking bad or breaking barriers? *North Carolina Journal of Law & Technology*, 18(5):244, 2017.

23. John Tromp. Cuckoo cycle: A memory-hard proof-of-work system. *IACR Cryptology ePrint Arch.*, 2014: 59, 2014.

24. Sunny King. Primecoin: Cryptocurrency with prime number proof-of-work. July 7th, 1(6), 2013.

25. Gavin Wood et al. Ethereum: A secure decentralised generalised transaction ledger. *Ethereum Project Yellow Paper*, 151(2014): 1–32, 2014.

26. V. Zamfir. Introducing Casper the friendly ghost, Ethereum blog, 2015.

27. Renato P. dos Santos et al. Consensus algorithms: A matter of complexity. In Melanie Swan, Jason Potts, Soichiro Takagi, Frank Witte and Paolo Tasca (eds) *Between Science and Economics*, pages 147–170. World Scientific, 2019.

28. Iddo Bentov, Ariel Gabizon, and Alex Mizrahi. Cryptocurrencies without proof of work. In *International Conference on Financial Cryptography and Data Security*, pages 142–157. Springer, 2016.

29. Aggelos Kiayias, Alexander Russell, Bernardo David, and Roman Oliynykov. Ouroboros: A provably secure proof-of-stake blockchain protocol. In *Annual International Cryptology Conference*, pages 357–388. Springer, 2017.

30. Anamika Chauhan, Om Prakash Malviya, Madhav Verma, and Tejinder Singh Mor. Blockchain and scalability. In *2018 IEEE International Conference on Software Quality, Reliability and Security Companion (QRS-C)*, pages 122–128. IEEE, 2018.

31. Mitar Milutinovic, Warren He, Howard Wu, and Maxinder Kanwal. Proof of luck: An efficient blockchain consensus protocol. In *Proceedings of the 1st Workshop on System Software for Trusted Execution*, pages 1–6. 2016.

32. Marko Vukolić. The quest for scalable blockchain fabric: Proof-of-work vs. BFT replication. In *International Workshop on Open Problems in Network Security*, pages 112–125. Springer, 2015.

33. Fahad Saleh. Blockchain without waste: Proof-of-stake. *The Review of Financial Studies*, 34(3): 1156–1190, 2021.

34. Maher Alharby and Aad Van Moorsel. Blockchain-based smart contracts: A systematic mapping study. *arXiv* preprint arXiv:1710.06372, 2017.

35. Harish Sukhwani, José M. Martínez, Xiaolin Chang, Kishor S. Trivedi, and Andy Rindos. Performance modeling of PBFT consensus process for permissioned blockchain network (hyperledger fabric). In *2017 IEEE 36th Symposium on Reliable Distributed Systems (SRDS)*, pages 253–255. IEEE, 2017.

36. Du Mingxiao, Ma Xiaofeng, Zhang Zhe, Wang Xiangwei, and Chen Qijun. A review on consensus algorithm of blockchain. In *2017 IEEE International Conference on Systems, Man, and Cybernetics (SMC)*, pages 2567–2572. IEEE, 2017.

37. Imran Bashir. *Mastering Blockchain: Deeper Insights into Decentralization, Cryptography, Bitcoin, and Popular Blockchain Frameworks*. Packt Publishing Limited, 2017.

38. Aleksey Novikov, Evgeny Gavrikov, Aleksandr Oleynik, Yuriy Zhirnov, and Nikolay Pestov. Blockchain technologies in managing socioeconomic systems: A study of legal practice. *Revista Inclusiones*, pages 452–461, 2020.

39. Tiana Laurence. *Blockchain for Dummies*. John Wiley & Sons, 2019.

18 Technology in Industry 4.0

Rashmy Moray and Amar Patnaik
Symbiosis Institute of Management Studies (SIMS),
Symbiosis International (Deemed) University

CONTENTS

18.1 INTRODUCTION

From the first to the fourth industrial revolution (IR), we have witnessed a substantial radical renaissance of technological evolution, from steam power-driven locomotives to electrical and digital mechanized output, making the production processes increasingly complex, automatic, challenging, and sustainable [1]. However, the aging population and intereconomic competition are some of the driving factors for the development of efficient technologies such as industrial Internet of things (IIoT)

DOI: 10.1201/9781003321149-18

and the cyber-physical systems (CPSs) to cater to the needs of effective systems and for improved industrial productivity [2].

Historians who studied the magnitude of radical transformations in the 18th and 19th centuries believed that industrial revolution should comprise and be reflected in the growth of macroeconomic aspects such as gross national product (GNP), gross domestic product (GDP), manufacturing produce, wealth creation, and production efficiency.

The manufacturing sector has been the driver of productivity growth by contributing 16% to the global GDP and playing a significant role in global economic development. The history of manufacturing has transformed dramatically, witnessing immense growth fueled by four big industrial revolutions.

18.2 EVOLUTION OF THE INDUSTRIAL REVOLUTIONS-FROM 1ST TO THE 4TH

18.2.1 INDUSTRY 1.0 | MECHANIZATION

The first industrial revolution (IR1) was characterized by "mechanization" and took place in the 18th century with the invention of the steam engine. It started around 1733 with the first cotton mill. Technological change was dominant in the years 1700–1830, and that period was deemed to have witnessed the IR1 by Thomas Ashton [3]. The period between the mid-1700s and mid-1800s represents the IR1, during which the production moved from being human centric to machine centric. It began in England and later spread across Europe and North America.

Britain recognized the significance of the inflow of technical know-how, striking the future of the 19th century in spreading technological knowledge and making microinventions their competitive advantage [4]. Powered by water and steam, organized production greatly increased the output of manufactured goods like cotton and textiles. In the late 19th century, Britain could not sustain the technological competitive advantage and lost to America due to the factors like R&D-based endogenous growth models, availability of human labor at a cheaper cost, and large markets [5,6].

The invention of the steam power-driven locomotive and the design and development of railways spurred industrial growth by way of providing cost-effective transportation of various materials and finished goods.

With these developments and inventions, the total factor of productivity improved marginally as compared to the pre-IR1 era, leading to substantial per-capita GDP growth. The pace and direction of technological advancement during the IR1 were fueled by the spirit of mechanical culture that brought science and technology together on the shop floor to achieve productivity [7].

From 1760s to 1770s, American colonies under British rule were not happy with the constant pressure of British imperial tax policies, and the lack of colonial representation led to repeated protests against and boycotting of British goods by American colonies, leading to the "American War of Independence" from 1775 to 1783.[1] Eventually, America won its independence from Britain in 1783.[2] However, after independence, owing to poor domestic production and shortage of labor force, increased dependency on Britain for import of goods compelled America to build a strong foundation on the indigenization of production of goods for a better economy [8]. Thus, with this vision, many technological inventions in the United States such as electricity, internal combustion (IC) engines, alloys and communication technologies contributed to the second industrial revolution (IR2) in the 19th century.[3]

18.2.2 INDUSTRY 2.0 | ELECTRIFICATION

The period from 1860 to 1914 is marked as the phase of the IR2, which, with the invention of electricity, witnessed a shift toward large-scale assembly lines with specialized labor and machines, powered by oil, gas, and electricity. It is popularly called the American Industrial Revolution that paved the path from a rural to an urban society. It began in the United States and later spread throughout many other parts of the world [8]. Samuel Slater is known as the "Father of the American Industrial Revolution."[4]

Electric motors, electric railways, and electric elevators are some of the key inventions of the IR2. American Industrialist Henry Ford, the founder of the Ford Motor Company, made a mass-produced Model-T on October 1, 1908, which reduced cost of production and helped Ford Motor make its automobiles available to the middle-class people, something that was not possible before. Thus, Ford became the most popular brand of choice for the Americans during this IR2. Development of fully interchangeable parts and laying the foundation for aviation industry by Wright Brothers in the United States, were some of the inventions that took place in this IR2. Invention of the telegraph and the telephone allowed the world to come closer with the exchange of information quickly over long distances and brought changes to communications technology. As the output and consistency of goods like automobiles, electric bulbs, and diesel engines increased massively, the real GDP of USA was more than seven times, with the real per-capita product more than double [8]. With the spread of plant production and output, equivalent growth in the cities was observed [9].

Increase in world population demanded more goods to be consumed, which led to the growth of global per-capita levels of industrialization. Factories, automobile vehicles, etc., burned fossil fuel to produce goods to meet the demand. Rise in fuel consumption, increased the level of CO_2 in the atmosphere which became the primary factor for global warming. This led to the rise in Earth's temperature causing an imbalance in global thermodynamics and thus, global warming and climate change [8], eventually leading to unforeseen natural disasters, and becoming threat to various actors involved in the ecosystem.

With this concern in mind, came the vision of moving toward a world with "Planetary Stewardship of Earth's Ecosystems" and the "Global Interconnectivity," which became the motivation for the Industry 3.0 [10].

18.2.3 INDUSTRY 3.0 | AUTOMATION

The third industrial revolution (IR3) is based on the convergence of communication and energy, shifting from using coal, oil, gas, and tar sands to using distributed energy sources such as the sun, wind, geothermal heat under the ground, biomass garbage, agricultural and forest waste, small hydro, ocean tides, and waves [11]. In the year 1960, the IR3 commenced with the usage of computers and paved the way through the expansion and conjunction of information and communications technology (ICT). For the first time, industrial robots were introduced by General Motors in 1961, and the first personal computers were seen during the 1980s. This period witnessed the "dotcom" boom fueled by investments in cyber-tech companies from around the globe [12].

Driven by technological advancement and the development of IT and automation, it transformed the manufacturing industries. Recognized as an era of high-level automation, PLCs (programmable logic controllers) and Industrial Robots, played an important role in the manufacturing of increasingly complex products, from automobiles with onboard computers to handphones. China, South Korea, Taiwan, Hong Kong, and Singapore—with their growing industrialization and being fueled by exports—were able to achieve higher economic growth during the IR3. One of the biggest inventions in this era was "3D printing," a technology that revolutionized the way the products were made and reduced the manufacturing time of complex parts. Thus, robotics and 3D printing are important transformational drivers in the IR3 [12].

Jeremy Rifkin, the architect of "The Third Industrial Revolution," described the five energy pillars as (i) shift to renewable energy, (ii) transformation of the building stock into green micropower plants to collect renewable energies onsite, (iii) distribution of hydrogen and supplementary storage technologies across infrastructures to store intermittent energies, (iv) the use of Internet technology to transform the power grid of every continent into an energy Internet that acts just like the Internet, and (v) transition of the transport fleet to electric plug-in and fuel-cell vehicles. Digitalization was set to mark the disruption of small, medium, and large manufacturers in this IR3 [13].

However, despite the technological innovation during these three IRs, emerging and advanced economies alike have experienced a decline in their manufacturing sectors, which became insignificant for national incomes and growth for the last couple of decades [14].

Apparently, the fourth industrial revolution (IR4.0) is the current mechanization of conventional engineering and industrial practices using contemporary smart technologies to bring the economy back to the growth trajectory [15]. The contemporary wave of I4.0 is characterized by "CPSs," which combines the power of ubiquitous ICT and physical manufacturing systems together to create a positive impact on productivity improvement.[5]

18.2.4   INDUSTRY 4.0 | CYBER-PHYSICAL SYSTEMS

The previous IRs used electronics and IT to automate the industry, but the Industry 4.0 (I4.0), is taking the shape of a digital resurgence in the industry [16] and is considered to be the extension of the IR3. It is trademarked by the combination of technologies that are bridging the gap between physical, digital, and biological spheres [17].

The key concepts of I4.0 were issued in the public domain for the first time as a part of a strategic initiative by the German government in the year 2011 and were incorporated in the "High-Tech Strategy 2020 Action Plan." The first three IRs represent Mechanization, Electrification, and Automation, respectively, followed by the I4.0 with the introduction of the IIoT, where the machineries, warehousing systems, and production facilities are poised to connect seamlessly through a cloud network, forming CPSs [18]. Thus, CPSs are the basis of IR4.0, with socio-technical interactions between the actors and the manufacturing resources [19].

While physical systems in manufacturing generate massive amounts of data, their connectivity to the digital world enables data collection and analysis to obtain meaningful insights for better visibility and faster decision-making. Thus, analytics is the key to manufacturers in the I4.0 [20].

The next level of automation and data exchange in manufacturing technologies, including CPSs, the IIoT, cloud computing, and cognitive computing, are some of the few key elements that define the "smart factories" in I4.0. Custom-based modularized production with personalized products, a service-oriented approach, a supply chain with near-real-time visibility, and a connected-ecosystem enabling customer-to-consumer interaction are some of the goals of I4.0 for providing value (quality and cost) and speed to the end customers.[6] Apparently, Industry 4.0 is evolving at an exponential rather than at a linear pace, disrupting each and every industry worldwide and heralding the transition of entire structures of production, management, and governance [16].

18.3   KEY TECHNOLOGIES IN INDUSTRY 4.0

As the I4.0 is characterized by the convergence of the physical and digital worlds, technology plays an important role in the transformation of businesses in the 21st century. Some of the key technologies that will drive I4.0 are briefly described below.

18.3.1   ROBOTICS

Developments in mechatronics, computing, and ICT have given birth to the contemporary field of robotics and autonomous systems (RAS) [21]. Known as collaborative robots (COBOTs), these smart robots interact with the environment through models and perform various tasks autonomously to optimize production efficiency and maximize industrial gains/profits [22]. Robots possess clear advantages in terms of speed, quality, and strength over manual labor. Industrial robots are used for material handling, which automates tedious, dull, repetitive, and often unsafe tasks at production lines. Enriching industrial robots with technologies like artificial intelligence (AI) and computer vision enables the completion of complex tasks.[6] With the help of AI, computer vision, and haptic-sensing, industrial robots will have capabilities similar to those of humans to interact, manipulate, train, and identify objects that are required to separately tailor the products. Thus, the new generation COBOTs would enhance the impact of intelligent automation by maximizing the abilities of both humans and machines together [21].

18.3.2 Industrial-IoT (IIoT)

IIoT refers to the technology and application of connecting devices and systems to exchange data and collaboratively implement services. It has been estimated that by 2030, over 50 billion IoT devices will be connected to the Internet.[7] In IIoT, sensors embedded in the machines continuously generate real-time data that provides manufacturing companies useful insights about the machines such as condition-monitoring, efficiency-analysis, and machine-utilization. Using appropriate services, networking technologies, applications, sensors, software, middleware, and storage systems, IIoT offers solutions and functions that develop insights to improve the capability of monitoring and controlling organizations' procedures and assets [23]. Additionally, through various interconnected devices and more centralized controllers, it provides scope for decentralized analytics and decision-making process to make real-time responses and reactions much faster [24]. This results in better availability and maintainability of enterprise assets, improved operational efficiency, faster time-to-market with reduced unplanned downtime, improved production efficiency, and exceptional levels of economic growth [25].

18.3.3 Artificial Intelligence (AI)

AI is the discipline of study in which machines, with the help of data from various industrial physical systems, IT systems and governed by certain algorithms, mimic human-like learning such as reasoning and self-correction within an interconnected digital ecosystem [26]. The subsets of AI comprising machine learning (ML) and deep learning (DL) are the technologies that help manufacturers with quality optimization, industrial automation, and adaptation capabilities of robots for improving their precision handling via tactile sensors [27].

ML is the ability of a machine to learn by itself without being explicitly programmed. Data, being the key requirement in ML, and collected from various sources of plants and machineries through IIoT technology, undergoes the process of data engineering to make it suitable for ML. Once the data are ready, with the help of relevant algorithms such as support vector machines (SVMs), k-nearest neighbor (KNN), and neural networks (NNs), the machine learns from the data and creates a model in the form of a relationship between the dependent and independent variables defined in the dataset. This model is then used for predicting the output from a given set of input parameters. This type of learning is often called statistical learning. Depending on the ML problems, there are three types of ML: supervised learning, unsupervised learning, and reinforcement learning. What a machine actually learns is derived from these three techniques of ML [28]. Machinery failure prediction, energy consumption optimization, and inventory optimization are some of the examples of ML use cases in manufacturing.[8]

DL is the next level of ML, which uses the NN architecture for learning. The simplest form of an artificial neural network (ANN) will have one input layer, two hidden layers, and one output layer. An ANN with more than one hidden layer is often called a deep neural network (DNN). DL involves DNN models with the NN, which has many layers such as an input layer, hidden layers, and output layers. The learning comes from these layers with built-in forward and backward propagation functions, which train the model to perform specific tasks as per the requirement [29]. One area of research in DNN is in the field of natural language processing (NLP), which uses two main types of architectures: convolutional neural network (CNN) and recurrent neural network (RNN) [30]. DL uses manufacturing data for product quality inspection, fault diagnosis, anomaly detection, etc. and offers insightful information to various stakeholders in manufacturing organizations, for making better decisions and manufacturing more productive [31].

Today, AI is considered to be the driving force behind I4.0 across both discrete and continuous manufacturing industries, thus giving rise to the intelligent factories where humans and CPSs interact in the cloud[9] with telecom technologies such as 2G, 3G, 4G, and the evolving 5G, as the basis for network communication.

18.3.4   5G NETWORK (5G)

5G is the fifth-generation mobile network, set to be the new global wireless standard. It has the potential to become the future communication platform of choice for many manufacturing industries, driving the future of I4.0 and smart manufacturing [32]. Ultrareliable low-latency communication, higher bandwidth, and support for higher device density, network slicing for virtual separation of networks, and high reliability are some of the key features that 5G provides the manufacturers to support their manufacturing mission-critical applications.[10] 5G is posed to set as the foundational platform for other I4.0 technologies such as AI, automation, augmented reality (AR), and IIoT.

The effect of 5G on I4.0 will be distinctive as its features provide solutions that are essential for manufacturing and helping manufacturers create fully integrated collaborative smart manufacturing systems that respond in the near-real time to meet the dynamics and environments in the plant and supply-chain networks and client requirements.[11]

18.3.5   CLOUD

Cloud is an infrastructure-as-a-service (IaaS) platform with computational capabilities such as storage, databases, servers, networking, software, analytics, and intelligence, which are offered as-a-service over-the-Internet or "the cloud." Cloud allows companies to tap into computing resources and services as a utility instead of building and maintaining them in-house [33]. It offers rapid innovation, flexible resources, and economies-of-scale for various business entities and individuals involved in the ecosystem of I4.0.[12] Today, multinational companies have globally distributed manufacturing setups and an increased requirement for data sharing for more production-related undertakings. As cloud technologies evolve over time, they will provide more opportunities to manufacturers for production system-related data-driven services by leveraging the functionality and machine data deployed on the cloud [34].

Cloud-based design and manufacturing (CBDM) is a term used to represent the process of product realization, which includes design and manufacturing resources into the model of cloud computing. Networked manufacturing, scalability, agility, ubiquitous access, multitenancy and virtualization, big data and IoT, IaaS, platform-as-a-service (PaaS), and software-as-a-service (SaaS) are some of the characteristics of CBDM [35].

As cloud services evolve, manufacturing companies around the world are looking to the cloud as a long-term investment in infrastructure and are the biggest adopters of hybrid cloud solutions (on-premises + cloud) to get a competitive edge in the market.[13]

Amazon Web Services, Microsoft Azure Cloud, and Google Cloud are the key cloud services being offered by the respective cloud players across the globe. Though cloud has got competitive advantages for the manufacturing industries, it poses challenges in terms of cyberattacks. This is where "cybersecurity" comes in as a technology discipline that plays an important role in protecting the plants and machinery-control systems from unauthorized access.

18.3.6 CYBERSECURITY

In the interconnected I4.0 realm, ensuring privacy and data security is the key to prevent cyber crime and cyber terrorism, which could disrupt the system and take it down when everyone is connected. This is no exception for manufacturing as it gets more digitized and adopts automation.[14] Thus, the existence of IIoT and the demand for the integration of information technology (IT) and operational technology (OT), has dramatically changed the entrance of cyber threats. Security threats and vulnerabilities of IIoT challenges are the vital reasons for cyber attacks [36]. Cybersecurity ensures to prevent such threats.

Vulnerabilities, threats, and attacks are the three vital concepts that form the core of cybersecurity problems. Vulnerability refers to any weakness that can be easily exploited within computing or networking systems that can lead to possible threats. A threat refers to any such process that can potentially infringe system's security policies that lead to an attack [37]. An attack is essentially contemplated to be an active process that intentionally pursues to breach a system's security policies. With the help of cybersecurity, organizations minimize the effect of attacks that potentially exploit vulnerabilities and reduce threats within systems to the extent possible [38].

According to the security software provider McAfee, cyber attacks targeting manufacturers increased sevenfold in 2020 between January and April, with financial losses caused by data breaches increasing 270% to $8.4 billion from January to March year on year.[15] This is the reason why cybersecurity in most recent times is considered to be one of the most popular and demanding technologies of research interest in the I4.0 world.

18.3.7 BIG DATA ANALYTICS

Big data refers to a technology of processing large volumes of structured, semi-structured, and unstructured complex data that gets generated during various manufacturing operational stages that traditional data processing methods cannot handle. It is characterized by three Vs, viz., volume, velocity, and variety, which differentiate it from traditional data [39]. It entails the process of the collection, transfer, storage, and analysis of data, for example, in predictive modeling, which supports decision-making.[16] Big data is essential in manufacturing to achieve efficiency gains and to uncover novel acumens to drive innovation. With the help of big data-based analytics, manufacturers can discover new data and identify patterns that allow them to enhance procedures, improve effectiveness of the supply chain, and identify variables that disturb manufacturing.[17]

Big data analytics are presently used for many industrial applications. This includes product lifecycle management [40], process redesigning [41], supply-chain management, and data analysis of the production systems for monitoring, detection of anomalies, root cause analysis, and additional knowledge [42].

Improved factory operations and production, product quality, supply chain efficiency, customer experience, and reduced machine downtime are some of the examples of use cases of big data analytics in manufacturing [43].

18.3.8 ADDITIVE MANUFACTURING

Additive manufacturing, in contrast to subtractive manufacturing, describes the process of building physical objects by adding materials rather than removing materials. It works by using 3D computer-aided design (CAD) models to create 3D-printed parts by depositing materials layer-by-layer with the help of a 3D printing machine onto a surface. Materials to make these parts can be of a wide variety of types, including metallic, ceramic, and polymeric materials, along with combinations in the form of composites, hybrid, or functionally graded materials. 3D printing is emerging as a valuable digital manufacturing technology of I4.0 [44]. Initially, it was only a rapid prototyping technology (RPT) that evolved over a period to advanced additive manufacturing technology, which fundamentally proposes an enormous opportunity for manufacturing, from tooling to mass customization across virtually all industries [45]. Additive manufacturing allows components to be stored in virtual inventories as 3D design files so that they can be manufactured on demand, a model known as distributed manufacturing. Such a decentralized approach reduces the cost of transport and simplifies inventory management by storing digital files instead of physical components.[18]

Some of the potential benefits associated with additive manufacturing are as follows: (i) direct translation of design to components, (ii) part generation with greater customization, (iii) ability to produce complex internal features, (iv) reduced weight of the parts, (v) substantial reduction in overall product development and manufacturing lead time, (vi) involvement of smaller manufacturing operations, (vii) high scalability, (viii) maximum material utilization with minimal waste, and (ix) capability to produce fully functional parts [46].

Although the complete adoption of this technology in an industrial setting requires reaching a certain level of design maturity, additive manufacturing has the potential to have a constructive impact on the industrial segment by minimizing production costs, logistics costs, inventory costs, and costs involved in the advancement and automation of new products [47].

18.3.9 Augmented Reality (AR)

In I4.0, as increasingly more data are generated and acquired from various sources, it is no longer unimaginable to think of a factory where not only everything is connected, but it is also viewable and interactive [48]. This is where AR comes in. AR is a technology that superimposes the virtual world onto the real world to present an enhanced customer experience [49]. AR helps users to visualize the real surroundings with the virtual environment embedded in them [50]. AR displays the supplement view of a real-world environment or objects with computer-generated inputs including sound, graphics, and GPS location. AR is used to enhance the decision-making capacity and efficiency of manufacturing stakeholders, typically with handheld displays or wearables like glasses.

The applications of AR in assembly, maintenance, and repair are found in manufacturing. Instructions are better and more effectively understood if they are obtained in the form of a superimposed physical-virtual experience rather than in the form of texts and images. Through AR, a maintenance person might see precisely which equipment needs servicing, including its potential issues, operation times, date of last service, and probable sites of failure [51].

18.3.10 Digital Twin (DT)

During the product design and development process, the product definition of a 3D model is created using a CAD software. Two types of data are mainly included in a 3D model: the geometric and non-geometric information stored in the specification tree. These data are then stored and managed in a product data management (PDM) software [52]. Once the 3D model is realized, the tooling design, production manufacturing processes, and even product function testing and verification process simulations and optimization are performed based on the model. Design functional proofing is done using computer-aided engineering (CAE), which simulates design performance, that is, whether or not design meets the functional criteria [53].

Once the design is ready and released for manufacturing and later realized physically, it then goes under testing, showing the actual behavior of the product functionality when subjected to real-world conditions. Results from the testing are then verified with the pass and fail criteria. If passed, the product moves to the production phase, and if fails, it undergoes the design modification until approved for passing criteria. The above process of product development has physical and digital world processes involved that are carried out separately. This gap of separation is bridged by the digital twin technology. It is the mapping technology for the complete lifespan process of physical systems in the virtual space. Thus, digital twin is a digital representation of physical products and the technology that allows the mirroring of real-time data between physical objects and digital models in the virtual world. Data from all aspects of the product's lifecycle are collated in the digital twin and analyzed for getting insightful information at every step of its lifecycle, such as the product design stage, production stage, and product service stage, which possibly could lead to performance improvements, productivity improvements, and improvement in customer satisfaction [53].

18.3.11 BLOCKCHAIN

Blockchain in manufacturing is a distributed digital record of transactions of individual records structured as "blocks", which are linked into a single list, or "chain", hence the name blockchain. It is also known as distributed ledger technology [54]. Each transaction added to a blockchain is validated with consensus from the stakeholders before being added to the blockchain [55].

Through a peer-to-peer network, blockchain enables manufacturers to streamline operations, gain greater visibility into supply chains, and track assets with a higher level of precision [56]. Blockchain serves the purpose of tackling the issue of possible inaccurate and tampered data, which poses the risk of huge threats to the interconnected networks of systems [57].

18.4 CONTEMPORARY SCENARIO

While it has been tough for several manufacturing enterprises to understand what I4.0 really is and that they are still in the process of evaluating its merits and demerits, the adoption of I4.0-related technologies in industrial manufacturing is gaining momentum with less than 30% adoption today and North American manufacturers having the largest adoption rates, leading to a huge gap of regional difference across nations [58]. At the same time, the adoption of I4.0 technologies by the automotive manufacturing industry at an early stage highlights a vast gap across industry adoptions. Enterprises like Siemens, General Electric, and Boeing are considered to be the leading manufacturers by adopting Industry 4.0.[19]

While countries all over the world such as the United States, Japan, China, the United Kingdom, Ireland, Sweden, and Austria have started adopting Industry 4.0 technologies, as per estimation the global I4.0 market is expected to reach 13,90,647 crore rupees by 2023 [59]. Also, according to a report published by the World Economic Forum, out of the 100 countries and economies included in the assessment, only 25 countries were found well positioned to benefit from I4.0.20 India, despite being the sixth-largest global manufacturing hub, is behind its global peers in I4.0 adoption. A substantial fraction of the Indian industrial segment is still in the postelectrification phase, with the technology usage restricted to the siloed systems that operate independently of each other. CPS, which is the basis of I4.0, is still at its budding stage [60]. Also, the micro, small and medium enterprises (MSMEs) segment is yet to adopt the automation technology, owing to its high cost of implementation [59].

However, the Government of India's vision and key focus on achieving its share of manufacturing target GDP of 25% by 2022 have offered an immense potential to the Indian Automobile Industry for possible expansion and embracement of the I4.0 technologies to create state-of-the-art manufacturing facilities in line with its global peers. The Automotive Mission Plan 2016–26 is one such initiative that depicts the government's combined vision to grow the automotive sector with technological maturity, global competitiveness, and a well-established institutional structure. The goal is to brand India among the top three benchmark locomotive industries across the globe and boost exports exponentially. With the combined effort of the Indian

Government and the Indian Automotive Industry, India is poised to create enabling ecosystem which will help India become a world-class manufacturing hub in the years to come [61]. With an investment from the Boeing Company and with a partnership with the Indian Institute of Science's (IISc) Centre for Product Design and Manufacturing, India's first smart factory in Bengaluru, equipped with data exchange in manufacturing and the Internet of Things (IoT) is being developed in line with the I4.0 standard framework [62].

However, the overall current scenario of I4.0 adoption readiness challenges in India can be due to (i) lack of understanding of the technologies, (ii) lack of an organized approach toward digital transformation, (iii) lack of interest to adopt I4.0 technologies, (iv) consideration of cheap labor in lieu of adopting automation, (v) a small volume of products making it not lucrative for investment, and (vi) skillset unavailability to adopt automation [59].

18.5 NEED AND SIGNIFICANCE

As customers' demands are increasing day by day, it becomes imperative for the manufacturers to be more agile and responsive to the dynamics of customer preferences by being mass-customization centric rather than being mass-production centric. To address this, I4.0 technology solutions provide manufacturers with innovative forms of personalization. Direct customer inputs can be considered during the product design stage, which will help establishments to progressively manufacture customized products with faster delivery times and lower costs [63–65].

The COVID-19 pandemic led to a sudden increase in the gap to connect to the manufacturing world, affecting operational visibility. Despite still trying to adjust to the new normal, the necessity of digital transformation in manufacturing organizations has become the key priority for business leaders. The value of I4.0 is suddenly recognized with the accelerated adoption of I4.0 technology solutions in advanced and developing economies as it is based on the vision of a fundamental process of innovation and digital transformation in industrial production.[20] Impacted by COVID-19, significant investments in digital transformation with an enhanced focus on I4.0 to improve productivity can be witnessed by the automotive industry future-proofing its technology estate.[21]

As I4.0 describes a fundamental innovation and transformation process for industrial value creation [19], it is poised to bring about manufacturing revolution all over the world, as the first three IRs did. This IR4.0 in the 21st century will be different as compared to its predecessors as the ecosystem will be more interconnected, much faster, and more responsive to the changing requirements. Information will flow throughout the value chain by connecting the producers and the consumers in this ecosystem. I4.0 will provide tremendous opportunities while creating the need for a highly trained, skilled, and flexible workforce and a production capacity that will answer the needs of tomorrow [61,66,67,68].

18.6 CONCLUSION

I4.0 is the convergence of the physical and digital worlds. Intelligent factories of I4.0 will have machines and products communicating with each other cooperatively, driving production in an efficient way to boost up productivity. Factories of I4.0 will have raw materials and machines interconnected with each other within an ecosystem of IoT. The vision of the I4.0 is to have smart factories with extremely flexible individualized and resource-friendly mass production resulting in tailor-made products at relatively reasonable prices. It also means highly flexible mass production that can be rapidly adapted to market variations because, in the future product lifecycles will be even shorter. Emerging technologies such as IIoT, cloud, big data analytics, and cybersecurity are pushing I4.0 forward, which promises to change the way business is done.

REFERENCES

1. D. Rodrik, "Premature deindustrialization," *Journal of Economic Growth,* vol. 21, no. 1, pp. 1–33, 2015.
2. J. Qin, Y. Liu, and R. Grosvenor, "A categorical framework of manufacturing for Industry 4.0 and beyond: changeable, agile, reconfigurable & virtual production," in *Procedia CIRP*, 2016.
3. N. Crafts, "The first industrial revolution: A Guided Tour for Growth Economists, The American Economic Review," in *Papers and Proceedings of the Hundredth and Eighth Annual Meeting of the American Economic Association*, San Francisco, CA, 1996.
4. N. F. R. Crafts and C. K. Harley, "Output growth and the British industrial revolution: a restatement of the Crafts-Harley," *Economic History Review*, vol. 45, no. 4, pp. 703–730, 1992.
5. N. F. R. Crafts, "Exogenous or endogenous growth? The industrial revolution reconsidered," *Journal of Economic History*, vol. 54, no. 4, pp. 745–772, 1995.
6. G. Nelson, "Rise and fall of American technological leadership: The postwar era in historical perspective," *Journal of Economic Literature*, vol. 34, no. 4, pp. 1931–1964, 1992.
7. M. Jacob, "Mechanical science of the factory floor," *History of Science*, vol. 45, no. 148, pp. 197–221, 2007.
8. H. K. Mohajan, "The second industrial revolution has brought modern social and economic developments," *American Institute of Science*, vol. 6, no. 1, pp. 1–14, 2020.
9. S. Kim, "Immigration, industrial revolution and urban growth in the United States," *1820 1920: Factor Endowments, Technology and Geography*, Working Paper 12900, 2007.
10. J. Rifkin, "Ecological, Ushering in a Smart Green Digital Global Economy to Address Climate Change and Create a More," *WordPress*, 2015.
11. T. Waghorn, "http://www.siaf.ch/files/130410-rifkin-2.pdf," 12 December 2011. [Online]. Available: http://www.siaf.ch/files/130410-rifkin-2.pdf
12. B. Roberts, "The third industrial revolution: Implications for planning cities and regions," 2015.
13. J. Rifkin, "The Third Industrial Revolution," *Spring*, pp. 32–34, 15 November 2012.
14. IMF, "https:// www.imf.org/B/media/Files/Publications/WEO/2019/October/English/text.ashx," 2019. [Online].
15. J. Taalbi, "Origins and pathways of innovation in the third industrial revolution," *Industrial and Corporate Change*, vol. 28, no. 5, pp. 1125–1148, 2019.
16. Schwab, "https://www.weforum.org/agenda/2016/01/the-fourth-industrial-revolution-what-it-means- and-how-to-respond/#," 14 January 2016. [Online]. Available: https://www.weforum.org.
17. J. M. Xu, "The fourth industrial revolution: opportunities and challenges," *International Journal of Financial Research*, vol. 9, no. 2, pp. 90–95, 2018.

18. H. Drath, "Industrie 4.0: hit or hype?," *IEEE Industrial Electronics Magazine*, vol. 8, no. 2, pp. 56–58, 2014.
19. W. Kagermann, "Securing the future of German manufacturing industry: recommendations for implementing the strategic initiative," Final report of the Industry 4.0, Working Group, European Commission, DG Employment, Social Affairs and Inclusion, 2013.
20. Y. M. Omar, M. Minoufekr, and P. Plapper, "Business analytics in manufacturing: Current trends, challenges and pathway to market leadership," *Operations Research Perspectives*, vol. 6, no. 1, pp. 100127, 2019.
21. L. Pawar, "https://www.ukras.org/wp-content/uploads/2018/10/UK_RAS_wp_manufacturing_web.pdf" 2016. [Online]. Available: https://www.ukras.org.
22. H. P. Karabegovic, "The role of service robots in Industry 4.0—smart automation of transport," *International Scientific Journal "Industry 4.0"*, vol. 4, no. 6, pp. 290–292, 2019.
23. S. Lampropoulos, "Internet of Things in the context of Industry 4.0: An overview," *International Journal of Entrepreneurial Knowledge*, vol. 7, no. 1, pp. 4–19, 2019.
24. D. B. Sen, "Smart factories: a review of situation, and recommendations to accelerate the evolution process," in *Proceedings of the International Symposium for Production Research*, 2018.
25. A. Gilchrist, "Designing Industrial Internet Systems," in *Industry 4.0: The Industrial Internet of Things*, Apress, pp. 87–118, 2016.
26. Z. Mosterman, "Industry 4.0 as a cyber-physical system study," *Software and Systems Modeling*, vol. 15, no. 1, pp. 17–29, 2015.
27. D. Lee, "Industrial artificial intelligence for industry 4.0-based manufacturing systems," *Manufacturing Letters*, vol. 18, pp. 20–23, 2018.
28. H. D. Wehle, "Machine learning, deep learning, and AI: What's the difference?," in *International Conference on Data Scientist Innovation Day*, Bruxelles, Belgium, 2017.
29. K. Atkeson, "The Transition to a New Economy after the Second Industrial Revolution," *Working Paper 606,* 2001.
30. W. Yin, K. Kann, M. Yu, and H. Schütze, "Comparative study of CNN and RNN for natural language processing," arXiv preprint arXiv:1702.01923, pp. 1–7, 2017.
31. Y. Wang, "Deep learning for smart manufacturing: methods and applications," *Journal of Manufacturing Systems*, vol. 48, pp. 144–156, 2018.
32. D. M. T. N. O'Connell, "Challenges associated with implementing 5G in manufacturing," *Telecom*, vol. 1, no. 1, pp. 48–67, 2020.
33. A. Fernández, S. del Río, V. López, A. Bawakid, M. J. del Jesus, J. M. Benítez, and F. Herrera, "Big Data with Cloud Computing: An insight on the computing environment, MapReduce, and programming frameworks," *WIRES Data Mining and Knowledge Discovery*, vol. 4, no. 5, pp. 380–409, August 2014.
34. M. Rüßmann, M. Lorenz, P. Gerbert, M. Waldner, P. Engel, M. Harnisch, and J. Justus, "https://www.bcg.com/publications/2015/ engineered_products_project_business_industry_4_future_productivity_growth_manufacturing_industries," 9 April 2015. [Online]. Available: https://www.bcg.com/.
35. D. Thames, "Software-defined cloud manufacturing for Industry 4.0," in *Procedia CIRP*, 2016.
36. B. Ervural, "Overview of cyber security in the Industry 4.0 Era," in *Managing the Digital Transformation,* 2018, pp. 267–284.
37. Hongmei He et al., "The security challenges in the IoT enabled cyber-physical systems and opportunities for evolutionary computing & other computational intelligence," in *IEEE Congress on Evolutionary Computation (CEC)*, 2016.
38. J. Lane Thames, "Distributed, collaborative and automated cybersecurity infrastructures for cloud-based design and manufacturing systems," *Cloud-Based Design and Manufacturing (CBDM)*, pp. 207–229, 2014.

39. B. Purcell, "Big data using cloud computing," *Journal of Technology Research*, vol. 5, no. 8, pp. 1–8, 2014.

40. F. Li, "Big Data in product lifecycle management," *The International Journal of Advanced Manufacturing Technology*, vol. 81, no. 14, p. 667 684, 2015.

41. K. N. Palma-Mendoza, "A business process re-design methodology to support supply chain integration: application in an airline MRO supply chain," *International Journal of Information Management*, vol. 35, pp. 620–631, 2015.

42. H. T. W. K. Ismail, "Manufacturing process data analysis pipelines: a requirements analysis and survey," *Journal of Big Data*, vol. 6, no. 1, pp. 1–26, 2015.

43. H.-N. Dai, H. Wang, G. Xu, J. Wan, and M. Imran, "Big Data analytics for manufacturing internet of things: opportunities, challenges and enabling technologies," *Enterprise Information Systems*, vol. 14, pp. 1–25, 2015.

44. U. M. Dilberoglu, B. Gharehpapagha, U. Yamana, and M. Dolen, "The role of additive manufacturing in the era of Industry 4.0," in *27th International Conference on Flexible Automation and Intelligent Manufacturing, FAIM2017*, Modena, Italy, 2015.

45. J. Butt, "Exploring the Interrelationship between Additive Manufacturing and Industry 4.0," *Designs*, vol. 4, no. 13, pp. 1–33, 2020.

46. S. A. M. Tofail, E. P. Koumoulos, A. Bandyopadhyay, S. Bose, L. O'Donoghue, and C. Charitidis, "Additive manufacturing: scientific and technological challenges, market uptake and opportunities.," *Materials Today*, vol. 21, no. 1, pp. 22–37., 2018.

47. R. Godina, I. Ribeiro, F. Matos, Bruna T. Ferreira, H. Carvalho, and P. Peças, "Impact assessment of additive manufacturing on sustainable business models in industry 4.0 context,". *Sustainability*, vol. 12, no. 17, p. 7066, 2020.

48. F. DePace, "Augmented reality in industry 4.0," *American Journal of Computer Science and Information Technology*, vol. 6, no. 01, pp. 1–7, 2018.

49. P. Fraga-Lamas, T. M. FernáNdez-CaraméS, Ó. Blanco-Novoa, and M. A. Vilar-Montesinos, "A review on industrial augmented reality systems for the industry 4.0 shipyard," 2018. doi: 10.1109/ACCESS.2018.2808326

50. V. Paelke, "Augmented reality in the smart factory: Supporting workers in an industry 4.0. environment.," in *Proceedings of the 2014 IEEE emerging technology and factory automation (ETFA)*, 2014.

51. C. Narcisa Deac, G. Calin Deac, C. Laurentiu Popa, M. Ghinea and C. Emil Cotet, "Using augmented reality in smart manufacturing," in *Annals of DAAAM & Proceedings*, 2017.

52. T. H.-J. Uhlemann, C. Lehmann, and R. Steinhilper, "The digital twin: Realizing the cyber-physical production system for industry 4.0," *Procedia Cirp*, vol. 61, pp. 335–340, 2017.

53. Z. Wang, "Digital twin technology, in *Industry 4.0—Impact on Intelligent Logistics and Manufacturing*," pp. 7–21, 2020.

54. T. Ko, J. Lee, and D. Ryu, "Blockchain technology and manufacturing industry: Real-time transparency and cost savings," *Sustainability 1*, vol. 10, no. 11, p. 4274, 2018.

55. S. Zhang and J.-H. Lee, "Analysis of the main consensus protocols of blockchain," *ICT Express*, vol. 6, pp. 93–97, 2019.

56. A. Vatankhah Barenji, Z. Li, W. M. Wang, G. Q. Huang, and D. A. Guerra-Zubiaga, "Blockchain-based ubiquitous manufacturing: A secure and reliable cyber-physical system," *International Journal of Production Research*, vol. 58, no. 7, pp. 2200–2221, 2020.

57. T. Kobzan, A. Biendarra, S. Schriegel, T. Herbst, T. Müller, and J. Jasperneite, "Utilizing blockchain technology in industrial manufacturing with the help of network simulation," in *2018 IEEE 16th International Conference on Industrial Inf*, 2018.

58. A. Schumacher, S. Erol, and W. Sihn, "A maturity model for assessing Industry 4.0 readiness and maturity of manufacturing enterprises," in *Procedia Cirp*, 2016.

59. K. Pundir, "The role of Industry 4.0 of small and medium enterprises in Uttar Pradesh," *ABS International Journal of Management*, vol. 6, no. 2, pp. 68–74, 2018.
60. V. V. Jadhav, R. Mahadeokar, and S. D. Bhoite, "The Fourth Industrial Revolution (I4. 0) in India: Challenges & Opportunities," *International Journal of Trend in Scientific Research and Development*, pp. 105–109, 2019. https://doi.org/10.31142/ijtsrd23076
61. S. Chouhan, P. Mehra, and A. Dasot, "India's readiness for Industry 4.0–A focus on automotive sector," *Confederation of Indian Industry (CII), Grant Thornton-An Instinct for Growth*, 2017.
62. A. Gautam, "Industry 4.0: The Industrial Revolution and New Concepts for the Factory of Future," *International Journal for Research in Applied Science and Engineering Technology (IJRASET)*, vol. 8, no. 6, pp. 1718–1722, 2020.
63. M. Wang, "Industry 4.0: a way from mass customization to mass personalization production," *Advances in Manufacturing*, vol. 5, no. 4, pp. 311–320, 2017.
64. J. Greenwood, "The IT revolution and the stock market," *Economic Review*, vol. 35, no. 2, pp. 2–13, 1999.
65. J. Greenwood, "The third industrial revolution: technology, productivity, and income equality," *Economic Review*, vol. 35, pp. 2–12, 2001.
66. G. Sarma, "Identities in the future Internet of Things," *Wireless Personal Communications*, vol. 49, no. 3, pp. 353–363, 2009.
67. L. R. D. D. Schaefer, "Distributed collaborative design and manufacture in the cloud—motivation, infrastructure, and education," in *American Society for Engineering Education Annual Conference Paper# AC2012–3017*, San Antonio, 2012.
68. H. Tuptuk, "Security of smart manufacturing systems," *Journal of Manufacturing Systems*, vol. 47, pp. 93–106, 2018.

NOTES

1 https://history.state.gov/milestones/17761783
2 https://en.wikipedia.org/wiki/American_Revolutionary_War
3 https://en.wikipedia.org/wiki/Second_Industrial_Revolution
4 https://www.pbs.org/wgbh/theymadeamerica/whomade/slater_hi.html8
5 https://www.i-scoop.eu/industry-4-0/
6 https://www.manufacturingtomorrow.com/article/2016/07/robots-in-manufacturing-applications/8333
7 https://www.statista.com/statistics/802690/worldwide-connected-devices-by-access-technology/
8 https://rapidminer.com/blog/6-ways-machine-learning-revolutionizing-manufacturing/
9 https://nexusintegra.io/artificial-intelligence-the-driving-force-behind-industry-4-0/
10 https://www.gsma.com/iot/wp-content/uploads/2020/04/202004_GSMA_SmartManufacturing_Insights_On_How_5G_IoT_Can_Transform_Industry.pdf
11 https://www.information-age.com/5g-is-the-heart-of-industry-4-0-123483152/
12 https://azure.microsoft.com/en-in/overview/what-is-cloud-computing/
13 https://www.loadspring.com/post/is-your-industry-fast-tracking-cloud-adoption/
14 https://www.euroscientist.com/modern-manufacturing-needs-cybersecurity
15 https://asia.nikkei.com/Business/Technology/Honda-and-other-smart-factories-fall-prey-to-hackers
16 https://www.sas.com/en_in/insights/big-data/what-is-big-data.html
17 https://www.manufacturing.net/operations/article/13228439/using-big-data-analytics-to-improve-production
18 https://amfg.ai/2019/03/28/industry-4-0-7-real-world-examples-of-digital-manufacturing-in-action/

19 https://iot-analytics.com/industry-4-0-adoption-2020-who-is-ahead/
20 https://www.mckinsey.com/business-functions/operations/our-insights/
    industry-40-reimagining-manufacturing-operations-after-covid-19
21 https://zinnov.com/covid19-and-the-automotive-industry-shifting-gears-on-the-other-
    side-of-coronavirus/

19 Intelligent Analytics in Cyber-Physical Systems

Manjushree Nayak, Priyanka P. Pratihari,
Sanjana Mahapatra, and Shyam Sundar Pradhan
NIST Institute of Science and Technology (Autonomous),

CONTENTS

19.1 INTRODUCTION

In past decades, companies focus on the quality products and improvement of their products. But now in today's world, enterprises are enforced to maintain competency and their position in the marketplace [1]. Recent technological advancements in computer science and technology have provided solutions to obtain and transfer colossal amounts of data from their fast-moving environment. Correspondingly, to handle such a vast amount of data "big data" came into account. Methods and terms such as the Internet of Things and interconnected systems are introduced to apply a solution to the "big data" environment [2].

The concept of total productive maintenance [3] evolved from an organization-centric focus on quality to a customer-centric focus on value creation and smart services. The evolution led to the development of prognostics and health management (PHM). PHM policies are designed to predict component failures, thus minimizing unexpected system downtimes. The underlying patterns are helping to avoid the costly failures and downtime of machinery. Near-zero breakdown can be obtained

DOI: 10.1201/9781003321149-19

by such a maintenance scheme. To make the process more consistent by adjusting and tuning processes, these invisibles visible can help.

Research in the field of intelligent maintenance has grown significantly and facilitated the development of intelligent maintenance systems. As computer and information systems progress, sensors, data collection equipment, wireless communication devices, and remote computing solutions become available. Using predictive analytics, such technologies are becoming the face of modern industries. An advanced predictive analytics system integrated with communication technology and machines is known as a cyber-physical system (CPS).

Since the concept came into account CPS has been the ever-expanding terminology in today's developing world. It integrates the physical systems with computational models. This includes process control, energy, transportation, medical devices, military, automation, smart structures, etc. An intelligent CPS is capable of providing self-awareness and self-maintenance. Implementing this predictive analysis with a decision-making system can give proper service with productivity. The CPS is the central hub of data management. A critical role is played at the fleet level.

19.2   CYBER-PHYSICAL SYSTEM

The CPS structure consists of five levels which are called 5c architecture [4]. This provides a guideline for CPS for various applications. This CPS structure consists of two main components, i.e., (i) streamlining data from physical space into cyberspace and receiving feedback from cyberspace through advanced connectivity and (ii) intelligent data analytics that helps to construct cyberspace. The 5c structure shows the workflow of the CPS system. The 5c structure of CPS consists of smart connections, data-to-info conversion, cyber, cognition, and configuration.

19.2.1   SMART CONNECTION

Connecting machines and their components to obtain accurate data is the first step to develop CPS for modern industries. Various devices or sensors are used to acquire a variety of data and store and transfer it to a central server. At this level, the workflow has a significant impact on the performance of CPS in the next levels, and it is possible to discover quality and accuracy through the system.

19.2.2   DATA-TO-INFORMATION CONVERSION

Analyzing and converting data into information is the core of the architecture. In recent years, algorithms and techniques for data mining have been developed. A variety of data sources, from machinery to business management data, will be used to implement such algorithms.

19.2.3   CYBER

The cyber level plays the role of a central information hub, which collects massive information from machines in machinery work. To extract additional data, specific analytics are used that provide a better understanding of individual machines within a fleet.

19.2.4 COGNITION

At this level, users decide by information provided to them through the presentation. Comparative information and individual machine status can make it easier to determine the priority of maintenance tasks.

19.2.5 CONFIGURATION

This level gives feedback from cyberspace to physical space. A supervisory level performs the self-configuration and self-adaptation of machines. By applying corrective and preventive decisions, it serves as a resilience control system.

19.3 INTERNET OF THINGS WITH THE ADVANCEMENT OF INDUSTRIES

IoT holds a very crucial role in building the foundation for the transformation of manufacturing using the latest technologies. It is also a communication platform for the CPS [5]. IoT consists of different interacting smart devices with IP addresses, which enables each device with a unique IP address.

IoT is very much capable of collecting, sorting, and merging different kind of data that comes from the different sources within the network. It provides an independent way of data management with the capability to get real-time status and access. IoT helps in the implementation of big data for the conversion of data to a piece of valuable information and then that to a knowledge which is then converted to action via a CPS structure.

The IoT platform can be effectively used for many applications in the CPS and some of the applications are customer-service management, resource planning, managing the total supply chain, etc. These all applications are only made possible due to the collection and categorization of those massive data that is being collected from different sources. IoT collects those data from the machines through the controller signals and from the sensors which are being installed within the network [6].

IoT helps in industrial informatics which includes the location and sensor and for data mining, and it tells us about the unlabeled data. The data that are collected instantly are sent to the platform and then it is analyzed to bring out the results in the form of tables, figures, and projected graphs. The enormous amount of data which are collected by IoT devices leads to the need for analysis which generally includes data mining, visualization, normalization, etc. The whole data abstraction process is made easier due to the help of the new embedded systems. Big data analytics has a five-layered architecture of a context intelligence platform out of which IoT is a layer that captures data from the IoT-enabled devices like sensor signals, geographical coordinates, etc. Out of the sensors, a very useful one is that which detects the temperature and intensity of light in that area. This sensor is important because a luminous environment affects the productivity of an industry. IoT provides many such innovative ideas to improve the productivity of the industries. IoT-based CPS also helps the service providers in many different ways. It is a kind of multitier cloud-based system with a hierarchical structure including the global and local devices [7].

IoT also plays an important role in industry 4.0 which takes industrialization to a different level. Industry 4.0 is capable of making production more efficient, sustainable, and flexible with the help of developed technologies. Often called the fourth industrial revolution, Industry 4.0 encompasses a wide range of technologies and devices and on this basis, Germany is leading others. Industry 4.0 also refers to the integration of industrial technologies with the internet to make productivity up to the need with better efficiency and adaptability [8]. Industry 4.0 with IoT is the future of industrial production [9]. Nine technologies help in the transformation of industry 4.0 and out of those nine, the industrial Internet of Things (IIoT) has an important role. The IIoT links all devices and components together with the sensing sensors to get data from them [10]. The technology of the internet provides us with a very better approach to the performance of the CPS [11]. Industry 4.0 is an integration of many technologies like data analysis, data mining, security of the network, cloud computing, and intelligence. These technologies help to increase the efficiency of production [12].

Industry 4.0 is mainly implemented in the industries to make the manufacturing sector digital. Due to these functionalities, the mobile data network or internet has become a compulsory element in the production sector. Machine-to-machine communication (M2M) is an integral part of IoT technology. It allows the exchange of info between CPSs [13]. Visual computing is a kind of technology that helps Industry 4.0 and IIoT systems to make their impact on the manufacturing process. Using Industry 4.0, it is possible to plan, configure, order, design, manufacture, and operate according to custom customer criteria. A vital component of the fourth-gen industrial revolution is the capability to offer good performance-based services for working on the data in these coming days [14].

19.4   METHODOLOGY FOR DESIGNING CPS-BASED INDUSTRY 4.0 SYSTEMS

Based on the capacities of digital actual frameworks, a better system can be developed for the applications of fourth generation industrial era in the process of CPS. In a previous segment, we discussed how inter-linkage provides access to information in a massive amount. In any case, sole accessibility of information does not make a critical benefit. Along these lines, a versatile yet strong approach is expected to make due, arrange and process information for additional investigation by PHM calculations. This technique must be adequately expansive to genuinely use all the benefits of digital actual frameworks. In this particular part, we bring forward "Time Machine Methodology for Digital Physical Systems". The purpose of this is to integrate accessible information in the ecosystem of the big data, to enable the calculations in PHM to take place efficiently.

Each and every part of the armada will make some agent Memories Machine record in the internet. This digital representative concentrates advantageous data from the pool of accessible information and standardizes it for additional investigation. Separated data incorporate yet not restricted to execution history, stress and burden, activity boundaries, framework designs, and support records. A fizzled genuine part should be eliminated from the parent device and it should not participate in testing again. The virtual representation of an entity which is also termed as digital-twin, will remain unaffected by time. These digital-twins bring about consistent collection of time machine records and subsequently assemble different activity boundaries from wide scope of indistinguishable parts. Standardization of boundaries, in which further examination endeavors must be led, guarantees the likeness of time machine records with one another for indistinguishable parts. Time machine records obey the various leveled connection of genuine parts also and each digital-twin has access to the records or data of its ancestor parts with progenitor parts. An environment so rich in data enriches PHM calculations and enables continuous anticipation with accurate observing the lines of production. At last, this philosophy brings extreme execution of digital physical framework right into it for planning an Industry 4.0 production line.

19.5 CASE STUDY

19.5.1 INDUSTRIAL ROBOT

This particular section provides a better way for the industrial robot health monitoring system. The main goal of this study was to design a method or approach for monitoring the health and condition of 30 industrial robots that are being used in the production process. By using torque and speed data to construct a robust prognostic and health management algorithm, we had to use a more complex multiregime approach due to the different speed ranges within a CPS. Monitoring torque has become an increasingly popular technique for detecting faults in industrial robots because of its nature, and most research efforts in this area are focused on determining torque. Furthermore, the nonlinear relationship between operating speed and torque makes determining the health of the robot a challenge for PHM algorithms. The model is used to determine the speed and torque of all robots as well as various settings like the ratio of the gears and loads, and also for the measure of pressure gauge, and the products or devices that are being assigned from the production part's allotted robots. These configuration parameters help the model standardize and adjust how data clusters provide a more accurate analysis of operations data. Consequently, every aspect of the study was performed on the cloud, where cloud platform was used to keep ta dataset securely, and the algorithms for the monitoring of health, were based on the details and those data which we already kept in our cloud system. Each instance's health was calculated using PHM algorithms. The user was presented with highly structured designs (graphs, charts, and diagrams) via a web interface (Figure 19.1).

19.5.2 VIRTUAL BATTERY

An electric or hybrid vehicle is not complete without a battery pack. In addition to concerns about battery safety, uncertainty about driving range, and reliability and

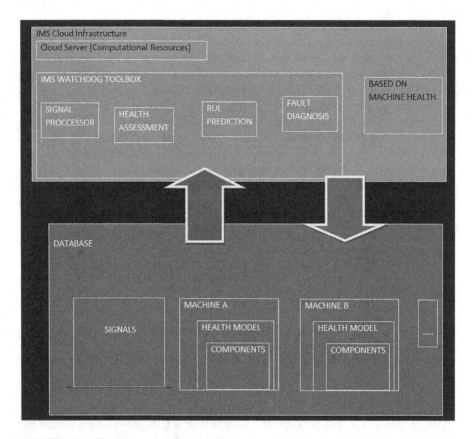

FIGURE 19.1 Industrial robot monitoring.

life of batteries, it is hard to achieve more widespread adoption of electric vehicles. Battery technology is becoming increasingly important to meet the demands of hybrid and electric vehicles have become increasingly popular in recent years due to challenges and rising demands [15]. Understanding battery performance requires a comprehensive understanding of its dynamics it is vital to know the conditions under several factors such as the environment temperature, humidity, driving style, charging level, and discharge level in rates and roads. A battery model, as a means for evaluating the health of a battery and predicting its failure in different conditions, will be necessary in order to achieve this understanding. With the help of such a battery model, suppliers can understand how changes in design and manufacturing processes impact battery performance and feed that information back to design and manufacturing partners [16]. An overview of an electric vehicle and battery health monitoring and prediction system is shown in Figure 19.2, including algorithms for calculating charge state and health, and classifying driving behavior. There is currently no battery model that considers the dynamics within a battery, but virtually all focus on individual cell models or battery packages. A battery's components, however, are interconnected. It is likely that interactions among all cells, conductors,

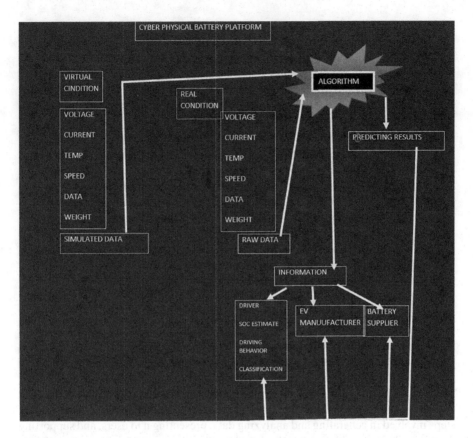

FIGURE 19.2 Virtual battery.

BMSs, and environment temperature affect battery performance significantly. Furthermore, manufacturing methodologies have a strong impact on the application conditions and behaviors of the battery.

In order to evaluate the impacts of multiregime changes in battery parameters and inputs on battery functionality, a simulated framework is required that is capable of executing the functions discussed above.

By combining new battery technology with intelligent and enabling tools, a "virtual battery" provides the following solutions:

1. By leveraging prognostic tools, health information can be transformed into useful information about battery performance, reliability, and readiness.
2. Delivering the right information at the right time through an adequate visualization system. The dashboard may display some information the driver needs to take immediate action, whereas other forms of information (such as more detailed diagnostic information) may be of more significance to maintenance and logistics personnel. Schedules maintenance or replacements at the center.

3. The tether-free communication enables real-time battery information to be gathered from vehicles and sent to a central logistics center, which can use GPS data to instruct the driver where to take the vehicle for service.
4. The ability to predict battery behavior based on external conditions (e.g., driving style, environment) and internal conditions (e.g., battery type and age) within new operations conditions, without having to make complex calculations.
5. Provide the designers with better estimates of the pack's lifespan expectation in the specific application as a result of a better hybrid cycle management.
6. Identify faulty packs and assist the battery manufacturers in refining their designs, manufacturing procedures, or raw material selections.
7. In addition to providing better vehicle monitoring during standard or extended warranty periods, wireless solutions could be integrated into prognostic tolls to detect faulty systems long before they need repair. The faulty system could be repaired prior to the usual inspection period. As a result, such instances may result in lower warranty costs. it may also be possible to integrate this system into engine management sensors.

19.6 CONCLUSION

This paper examined current trends regarding the implementation of CPSs in manufacturing industry. A generic architecture for implementing CPS in manufacturing is necessary since industrial big data has become a challenge for factories. In this article, we discuss the 5C architecture for automating and centralizing data processing, health assessment, and prognostics. This architecture covers all the steps involved in generating and analyzing data, presenting it to users, and supporting decisions. It is possible to use the health information generating by a system to perform higher-level functions such as maintenance scheduling and improve control to enhance overall system productivity. Using a case study of CNS saw machines, a brief look at the capabilities of the 5C architecture is demonstrated. The case study shows the application of the 5C architecture in a manufacturing environment for the processing and management of a fleet of CNC saw machines. In the current state of integration, the 5C CPS architecture is in its infancy. Advancements in all five levels of the architecture are therefore realistic. A distributed data management system with new algorithms for fleet-level analysis of machine performance has significant potential for advancement at the cyber level.

REFERENCES

1. L. Da Xu (2011) "Enterprise Systems: State-of-the-Art and Future Trends." *Industrial Informatics.*
2. E. Dumbill (2013) "Big Data Is Rocket Fuel." *Big Data.*
3. Big Data in Cyber-Physical Systems, Digital Manufacturing and Industry 4.0 Lidong Wang, Guanghui Wang.
4. Industrial Big Data Analytics and Cyber-Physical Systems for Future Maintenance & Service Innovation Jay Lee, Hossein Davari Ardakani, Shanhu Yang, Behrad Bagheri.

5. Research on IoT Based Cyber Physical System for Industrial Big Data Analytics C.K.M. Lee, C.L. Yeung, M.N. Cheng.
6. A. Stork (2015) Visual Computing Challenges of Advanced Manufacturing and Industrie 4.0. *IEEE Computer Graphics and Applications*, March/April 21–25.
7. M. Anne and M. Gobble (2014) News and Analysis of the Global Innovation Scene. *Research Technology Management*, November/December, 2–3.
8. Audio-Tech Business Book Summaries, Inc. (2015) Industry 4.0 and the U.S. Manufacturing Renaissance. *Trends E-Magazine*, June, 4–10.
9. I. R. Anderl (2014) Industrie 4.0- Advanced Engineering of Smart Products and Smart Production, Technological Innovations in the Product Development. *19th International Seminar on High Technology*, Piracicaba, Brasil, October 9th, 2014, 1–14.
10. I. Singh, N. Al-Mutawaly, and T. Wanyama (2015) Teaching Network Technologies That Support Industry 4.0. *Proc. 2015 Canadian Engineering Education Association (CEEA15) Conf., CEEA15*; Paper 119, McMaster University; May 31–June 3, 1–5.
11. J. Bechtold, C. Lauenstein, A. Kern, and L. Bernhofer (2014) Industry 4.0-The Capgemini Consulting View. Technical Report, Capgemini Consulting.
12. R. Davies (2015). Industry 4.0 Digitalization for Productivity and Growth, European Parliamentary Research Service (EPRS), Briefing, 1–10.
13. J. Lee, B. Bagheri, and H. A. Kao. (2015) "A Cyber-Physical Systems Architecture for Industry 4.0-Based Manufacturing Systems." *Manufacturing Letters*, 3: 18–23.
14. F. T. S. Chan, H. C. W. Lau, R. W. L. Ip, H. K. Chan, and S. Kong. (2005) "Implementation of Total Productive Maintenance: A Case Study." *International Journal of Production Economics*, 95(1): 71–94.
15. F. Ju, J. Wang, J. Li, and G. Xiao. (15 January 2013) "Virtual Battery: A Battery Simulation Framework for Electric Vehicles." *IEEE Transactions on Automation Science and Engineering*, 10(1): 5.
16. Y. He, W. Liu, and B. J. Koch. (2010) "Battery Algorithm Verification and Development Using Hardware-in-the-Loop Testing." *Journal of Power Sources*, 195: 2969–2974.

20 An Overlook on Security Challenges in Industry 4.0

R. Ramya, S. Sharmila Devi, and Y. Adline Jancy
Sri Ramakrishna Engineering College

CONTENTS

20.1 INTRODUCTION TO INDUSTRIAL INTERNET OF THINGS

The Internet has modified the manner humans communicate, do commercial enterprise, and collaborate. Now, the point of interest has switched to attaining the identical outcomes for machines. Systems builders have spent the previous couple of years that specialize in integrating sensors, facet nodes, and analytics to create clever structures that flip operations into high-productiveness settings. The industrial Internet of things (IIoT) is made up of those interconnected structures [1]. This is the maximum disruptive business revolution in history, touching regions as numerous as healthcare, strength, transportation, and production. Engineers in each enterprise will discover a manner to make use of the brand new abilities created via way of means of connecting system and methods to more effective compute and analytics abilities within side the coming years. The IIoT is the extension and use of the IoT

DOI: 10.1201/9781003321149-20

in business sectors and packages [2]. By that specializing in system-to-system (S2S) connectivity, huge statistics, and system learning, the IIoT allows industries and groups to grow their performance and dependability of their operations. The IIoT consists of business packages along with robotics, clinical gadgets, and software program-described production methods.

The IIoT is going past the conventional customer electronics and bodily tool inter-networking that the IoT is understood for. It is prominent via way of means of the mixing of statistics generation and operational generation. Operational methods and business management structures are networked with human–machine interfaces (HMIs), supervisory control and data acquisition (SCADA) structures, distributed control systems (DCSs), and programmable logic controllers (PLCs). Industry can take advantage of more machine integration in phrases of automation and optimization, in addition to better visibility of the delivery chain and logistics, way to the convergence of information technology (IT) and operational technology (OT). Smart sensors and actuators make it less complicated in capturing data and make robust infrastructure in industries, extending the same to other domains like agriculture, healthcare, production, and transportation.

As a part of the fourth business revolution, referred to as Industry 4.0 [3], the IIoT is essential to how cyber body structures and manufacturing methods will adapt with using huge statistics and analytics. Industrial system and infrastructures use actual-time statistics from sensors and different reassets to resource in "decision-making," permitting them to produce insights and carry out unique actions. The IIoT is vital in a broader feel to be used instances regarding networked ecosystems or surrounds, along with how towns and factories grow to be clever towns and clever factories. The normal series and transmission of statistics through clever gadgets and systems present many possibilities for sectors and groups to grow. The IIoT also can be used to optimize asset use, expect failure locations, or even set off upkeep sports on their own. Businesses can also additionally acquire and examine large quantities of statistics at quicker speeds via way of means of the usage of linked and clever gadgets. This will assist to bridge the space among manufacturing and standard offices, in addition to growth scalability and performance.

IIoT is a subset of sensors, automation, machine-to-machine (M2M), embedded systems, and industrial controls. They follow IT protocols rather than OT protocols where personnel is trained in IT. Connected with trillions of devices, IIoT has an umbrella of nontraditional systems with a combo of standardless protocols, remote management, and authorized/nonauthorized devices. Evaluation of variable environments in IIoT leads to security breaches. The chapter focuses on various attacks faced by target entities like devices, DCSs, and communication networks (Figure 20.1).

20.1.1 Security Troubles and Issues in Enforcing the IIoT

The adoption of business IoT technology has significantly benefited companies within side the production and transportation sectors (IIoT). Since IIoT structures have grown to be extensively used, significantly in safety, formerly insurmountable

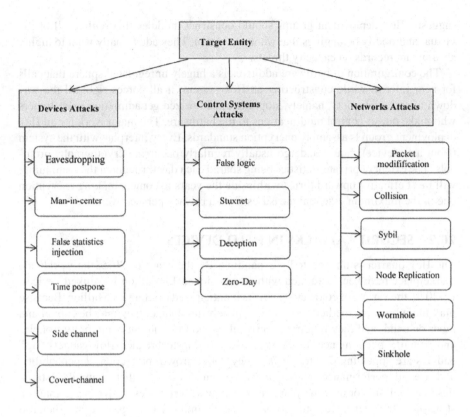

FIGURE 20.1   Various security attacks in IIoT.

technical problems were without problems overcome. Predictive protection, for instance, can assist a water remedy commercial enterprise with the aid of using placing sensors on its machinery. The sensors acquire overall performance data, examine it, and flag ability dangers, permitting the commercial enterprise to assume protection desires and run extra effectively. Despite the plain tangible welfare of extra virtual connectivity, IIoT deployment does offer a few IoT safety worries and challenges.

To sell industry-huge standards, the World Economic Forum (WEF) has prepared an insightful observation on IIoT protection and safety fine practices. Here are four IoT safety-demanding situations noted within side the notice that producers should grapple with. One of the maximum large issues [4] is while integrity legacy software program with IIoT systems is a loss of maintenance. While checking out legacy software program is one challenge, it is far vital that IIoT gadgets (and related software programs) are examined carefully in the course of their lifespan. Methodologies need to include unit, system, acceptance, regression checking out, and danger modeling, together with retaining a stock of the supply for any 1/3 party/ open supply code and additives utilized. As greater gadgets hook up with a shared community, it will become tougher for IT groups to discover risks. As the WEF suggests, IIoT deployment groups should construct modules that outline all of the virtual and bodily belongings that want protection. They additionally want to higher accumulate records on capacity threats.

The configuration of hardware additives is a hugely unregulated sphere that calls for near interest while constructing an IIoT system. It all comes right all the way down to endpoint safety, namely, stopping unauthorized get admission to customers who make out-of-control modifications to the hardware. The major work for an IIoT deployment group is assembly encryption standards. Every interplay with the system (from any device) should undergo usually regularly occurring cryptography proto-cols. The extent of private statistics being shared each day throughout the community will most effective upward thrust withinside the years to come – making encryption one of the maximum essential measures from a privacy perspective.

20.2   SECURITY ATTACKS IN FIELD DEVICES

The IIoT quarter is unexpectedly expanding. By the cease of 2015, the worldwide marketplace have been saturated with an envisioned 134 billion IoT-linked devices. By 2020, the variety of connected gadgets is anticipated to attain 38.5 billion. Because maximum business methods are not completely automated, they may be capital and labor demand, and they require a powerful method to find out community troubles and beautify performance, many groups are trying to broaden IIoT gadgets. IIoT gadgets are appealing due to the fact they could growth protection, dependability, and strength performance. These gadgets encompass sensors that accumulate statis-tics in actual time or over the years and connect with statistics analytics and manage structures. In order for utilities to prioritize community upkeep, it is far crucial to examine statistics traits on the way to supply actionable insights. Data styles should be analyzed so as for utilities to offer significant insights and prioritize community enhancements or screen delivery call for mismatches.

To lay out protection, it is far vital to apprehend from which the capability threats arise. Attacks may be released over all community links: via employer connections, connections via different networks on the manage community layer, and/or connections at the sector tool degree [5]. The Stuxnet assault may also had been the riding aspect that made protection researchers aware of vulnerabilities in area gadgets themselves. According to many sources, Stuxnet's goal changed into an Iranian uranium enrichment plant. Its purpose changed into to adjust this sys-tem jogging on particular PLCs to alternate centrifuge rotor speeds, inflicting harm to the rotors, at the same time as reporting fake values to cover the adjustments from machine operators. Stuxnet used more than one exploit to unfold to computer systems over networks and via thumb drives. Once Stuxnet inflamed a pc jogging a Siemens PLC configuration tool, it checked the configuration to peer if it had reached its goal, then changed the good judgment programming on particular PLCs in that machine. The complexity of the assault changed into wonderful as it required more than one skillsets, took a huge quantity of resources, and changed into very centered and self-limiting. Although the Stuxnet assault changed into extraordinarily com-plex, the real subversion of the PLCs changed into made trustworthy with the aid of using their design.

Some of the usual place assaults are as follows:

- **Eavesdropping assault**

  By tracking the community, an adversary can benefit from touchy statistics approximately the behavior of the community to perpetrate similar assaults. Network site visitors' analysis, even of encrypted packets, can screen statistics and compromise privates.
- **Man-in-the-center assault**

  In this assault, the adversary sits among speaking gadgets and relays conversation among them.
- **False statistics injection assault**

  A fake injection assault is a deception assault, wherein the adversary injects fake statistics into the community; for example, via way of means of sending malicious instructions on a subject bus.
- **Time postpone assault**

  An attacker injects more time delays into measurements and manages values of the structures, which could disturb balance of the machine and reason system to crash.
- **Replay assault**

  In a replay assault, valid packets may be retransmitted via way of means of an adversary. This can appear in numerous ways: a true however compromised node may want to ship the statistics.
- **Spoofing assault**

  Spoofing assaults are in which an attacker's node impersonates a machine entity. A loss of good enough authentication management mechanisms method that entities can masquerade as each other via way of means of falsifying their identification to benefit illegitimate get entry to.
- **Side channel assaults**

  Side channel assaults are done the usage of numerous strategies that examine statistics leakage from hardware and software program along with analyzing strength consumption, mild emissions, optical signal, site visitors flow, timings, electromagnetic, acoustic, and thermal emission from hardware additives and faults that arise with inside the machine [6].
- Covert-channel assaults

  This is an assault that uses a compromised tool and valid conversation channels to leak touchy statistics out of a stable environment, bypassing security features.

20.2.1 Securing an IIoT Device

Ayyeka, as an IIoT commercial enterprise, created a device that consists of present-day cybersecurity great practices. The essential ranges to shielding IIoT gadgets and recognizing tool vulnerabilities may be protected on this article. Embedding cybersecurity into IT layout, safeguarding conversation protocols, and integrating extra security features for offering statistics to SCADA structures are only some examples. Installing a firewall and an intrusion detection machine is the maximum fundamental degree of safety intrusion detection systems (IDS). Firewalls, on the alternative hand, may be set to pick out while a suspicious man or woman is coming near their property. As safety toward unauthorized intruders, a firewall is managed via way of means of protection guidelines that adjust incoming and outgoing community site visitors.

Encrypting sensitive statistics is the subsequent step. Hackers are nearly positive to try and intercept conversation from subject gadgets with inside the application enterprise. As a result, it is crucial to assume like a hacker and find out the community's weakest hyperlink in order that extra security features can be installed place. For example, all subject tool communications are encrypted with the SCADA machine.

20.2.2 A Step-by-Step Security Strategy for IIoT Devices

To assist corporations to plan their business virtual transformation correctly and securely, AWS recommends a multilayered method to stable the industrial control systems (ICS)/OT, IIoT, and cloud environments, which are captured in the following steps [7]:

- Conduct a cyber-safety threat evaluation of a not unusual place framework (which includes MITRE ATT&CK) and use it to tell gadget design.
- Maintain an asset stock of all related properties and updated community architecture.
- Provision cutting-edge IIoT gadgets and structures with specific identities and credentials, practice authentication and get admission to manipulate mechanisms.
- Prioritize and put into effect OT and IIoT precise patch control and outline suitable replacement mechanisms for software program and firmware updates.
- Secure production facts at the threshold and within the cloud through encrypting facts at relaxation and create mechanisms for stable facts sharing, governance, and sovereignty.

20.3 SECURITY BREACHES IN INDUSTRIAL CONTROL SYSTEMS

In today's Internet-connected society, concerns about more prominent cyber attacks. Multiple devices, computers, controllers, communications channels, and software are integrated to communicate and operate industrial processes in the IoT. As a result, both inside and outside the communication network, the IoT is vulnerable to physical intrusion and viruses. A hacker with awareness of software, process equipment, and networks can employ electronic means to really get access to IoT if it is not protected.

The main reasons most industry devices get hacked are as follows [8]:

a. Many plants' devices go weeks or months without receiving any security upgrades or antivirus software.
b. Because many of the controllers used in IoT networks were created in an era when cybersecurity was not a priority, they can be disturbed by malformed network traffic or even by large volumes of correctly formed traffic.

c. Many IoT networks have several points of entry for cybersecurity threats, allowing them to overcome existing security measures. Laptops brought into and out of facilities, as well as USB sticks used by several PCs without being properly scanned for malware, are common examples.

d. Many IoT networks are still built as a single, big, flat network with no physical or virtual protection between networks. This aids in the rapid spread of malware, even to remote locations devices.

20.3.1 ATTACKS IN ICS SYSTEMS

Zero-day attack happened in SCADA where the nature of attack is not predicted [9]. False logic attack refers to giving false command to SCADA and wrong data is obtained by ICS [10]. Deception is another attack on SCSDA and DCS where wrong commands are issued. The operators are subjected to reconnected mean while the hacker issues the incorrect command to the system. The Stuxnet attack is a major attack on PLC leading to misbehavior of it and causing the controller to reprogram [11].

20.4   SECURITY CHALLENGES IN COMMUNICATION NETWORKS

The IoT integrates everything on the Internet and in the telecommunications network. But it also introduces a special set of challenges for the development of technology due to the recent revolution in communications and computers. Signaling, security, authorization, open ports, and encryption are some of the main security issues affecting IoT connectivity, as follows [12]:

a. **Signaling:** bidirectional strong signal is necessary when IoT devices are connected in order to collect and route data between smart devices effectively, as the acquired data must move fast from point A to point B. Manufacturing message specification (MMS) packet modification attack is done while communication between SCADA and control center [13].

b. **Security**: developers must come up with some innovative solutions because the majority of smart devices do not have the basic necessities of hacker security. Security measures should offer enough privacy and security for end users for a constant flow of data moves between devices, networks, and gateways when millions of smart devices are interconnected. Collision attack leads to rearrange the checksum in the message frame. Fake identities of nodes are generated in Sybil attack. Node replication happens by cloning the sent nodes [14].

c. **Authorization:** as there are more connected devices, the IoT will produce and store an increasing amount of personal data, making it more attractive to hackers and increasing the risk of security breaches. We must confirm that the server or smart device being used has the appropriate authorization to send and receive data before sending or receiving any data.

d. **Open ports:** the issue is that bidirectional connection is required in order to send and receive data, but open ports must not exist because an open port exposes a system open to attack. Wormhole attack makes the node to be in a perception of it is in the right routing direction. Sinkhole attack makes to transmit fake data to neighbor node [14].

e. **Encryption:** The use of encrypted data transmission services by 70% of smart devices makes personal information accessible. It requires end-to-end encryption between our smart devices and the servers being used in order to send and receive any data secure manner.

Although there is no dispute that the IoT is transforming industries and improving everyone's consumer experience, security and data misuse are perennial concerns. There will be circumstances at which data may be misused because there is so much key information traveling around. The industries are still implementing the IoT, and it will take some time before users begin to notice and explore with the changes. The IoT concept relies on device communication, and smart things can communicate with one another using network technology.

20.5 NECESSITY OF DATA PROTECTION

The recent expansion of information and communication technology (ICT) has offered significant economic and technological prospects. Meanwhile, this has become the hornet's nest of cybercrime, allowing for the theft of personal information and identities, further more eavesdropping on people's professional and private life and other connected difficulties. Another significant challenge is that IoT, which links a multitude of devices for a two-way flow of information, and captures and archives zillions of personal data that can be exploited or misused by third parties.

20.5.1 INDUSTRY 4.0 CYBERSECURITY

The handling of personal data belonging to employees and consumers poses the greatest danger in a smart factory. Personal data may be captured, analyzed, and in certain cases transmitted combined with other data in any scenario of networking between humans, manufacturing unit, and logistical systems. Pellucid for employees and consumers about the technology being used, as well as appropriately qualified data protection officers who monitor data acquisition and processing in the context of Industry 4.0, are essential first steps toward a data-protection-compliant embrace of Industry 4.0.

20.5.2 IIoT AND BIG DATA ANALYTICS (BDA)

Enormous volumes of data are generated by IIoT systems. In industrial systems, data are generated through smart sensor networks, devices, log files, and other sources. Inevitably, with so many distinct and varied data sources, it is likely that the data will be highly diverse and structured, unstructured, or both. Volume, reliability, velocity, diversity, validity, and volatility are all characteristics of big data. As a consequence of the partnership with the IoT and cyber-physical systems (CPS), these production systems generate a tremendous amount of data, a wide diversity of data, and a high level of veracity. Metadata are collected from manufacturers, vendors, product designers, clientele, and consumers for BDA. Despite its unrivaled significance in furthering Industry 4.0, big data's distributed structure poses a number of issues. Data security, privacy, and access control, as well as data storage, are among the most important concerns.

Manufacturers can use BDA in manufacturing to discover the most up-to-date information and recognize patterns, allowing them to improve processes, regulate supply chain, and identify variables that influence production. According to reports, BDA in the manufacturing industry would approach $4.55 billion by 2025 [15]. Intelligent manufacturing ship floors are constructed using BDA, which represents tangible Internet-based logistics data. To collect data, they used radio-frequency identification readers, tags, and wireless communication networks on the factory floors. They used this information to create a visualization of the logistics chain and to assess the efficiency of logistics operators and operations [16].

20.5.3 IoT Data Privacy Leaking

Most businesses are increasingly aware of the importance of external security threats. As a result, we create a security system and a new security solution, as well as handle the outside security. Internal security concerns have become increasingly relevant as a result of recent insider data leaks and system sabotage. Data ingress by internal personnel is 50%, and data leakage by executive officers is up to 20%, reported on a study done by Websens Security Labs. This study demonstrates that only the outflow is no longer significant. Finally, the significance of user behavior analysis (UBA) technology is highlighted [17].

Many companies are investing on UBA and a few are narrated here. IBM uses analytics-based intelligence, and Netcruz focused on a combined platform of finding external infringement and internal information control. Through aberrant traffic categorization and traffic source inquiry, export management company (EMC) developed network forensics that can detect and respond to sophisticated assaults [18].

Typical raw data storage, energy limitations, encryption limitations, lack of standardized structure, and a few more factors all contribute to privacy leaks [19,20]. Data and intellectual property are disseminated across supply chains and stakeholders with Industry 4.0. Consumers and suppliers are integrating their systems. Data are dispersed among the systems, implying a broader security scope. As the number of stakeholders grows, so does the number of entry points that are subject to assault. Another assault vector is the convergence of information and operational technology. This becomes harder to develop an end-to-end security system.

20.5.4 Standard Security Solutions

Simple, well-known solutions, such as standard IT systems, access authentication, access control, and key management, may help to lessen the likelihood of a breach. Application protection, antimalware, and antivirus are a few examples of standard information security techniques. Network consumption, traffic, and logs may all be tracked, as well as network segmentation. Personnel security training is also a viable option. Patching methods could be useful for keeping industrial systems current and secure. Regular system audits and penetration testing are also important safeguards.

20.5.5 OPPORTUNITIES TO OVERCOME BDA CHALLENGES

An analytic team is to be recruited to frame strategies on data analysis in manufacturing companies. Standards need to be introduced in data collection, storage, and transmission from the floor or site. As data obtained from field devices are unstructured, a synchronized and structured database needs to be established. Data storage becomes costly as the data are expensive, and as the data processing algorithms works on large data size, size reduction techniques are being encouraged. A backup system for data is required when there is a possibility of negative BDA analysis. Finally, standards need to be improved in legislation policies and information systems [21].

20.6 CONCLUSION

In this chapter, the primary aim was to explore the current state of the art of security challenges in Industry 4.0 relating technologies in the manufacturing and automation industry. In Industry 4.0, industrial revolution is happening and it is accomplished with smart in production, logistics, storage, maintenance, and control system. On the other hand, there these smarter things are data hungry and prone to security attacks. As IIoT is a combination of field devices, control devices, and communication networks, attackers can target on any of the mentioned areas. Hackers targeting the field devices like sensors or industrial machines leads to hindrance in receiving data from the field. Different ICS system attacks are discussed in this chapter. Usage of dynamic IP for SCADA gives a better solution for ICS attack. In networks attack, wireless media are more prone compared to wired as the hacker can falsely route or trap data packets. The final chapter concludes the necessity of data protection, as IIoT are used in data-driven applications. Industry 4.0 extends its wings to BDA and cloud storage, and this tells the security breaches in data storage and analysis. UBA is implemented in leading industries as internal data leakage overcomes the external data hack.

REFERENCES

1. Matthew N. O. Sadiku, Yonghui Wang, Suxia Cui, and Sarhan M. Musa, "The industrial internet of things," *IJASRE*, vol. 3, no. 11, pp. 1–5, Dec. 2017.
2. A. Gilchrist, *Industry 4.0: The Industrial Internet of Things*. Springer, 2016.
3. Adam Mentsiev, Elina Guzueva, and Tamirlan Magomaev, "Security challenges of the Industry 4.0," *Journal of Physics: Conference Series*, vol. 1515, 032074, 2020.
4. S. R. Chhetri, N. Rashid, S. Faezi, and M. A. Al Faruque, "Security trends and advances in manufacturing systems in the era of industry 4.0," *2017 IEEE/ACM International Conference on Computer-Aided Design (ICCAD)*, 2017, pp. 1039–1046.
5. Nilufer Tuptuk, and Stephen Hailes, "Security of smart manufacturing systems," *Journal of Manufacturing Systems*, vol. 47, pp. 93–106, 2018.
6. M. A. Al Faruque, S. R. Chhetri, A. Canedo, and J. Wan, "Acoustic side-channel attacks on additive manufacturing systems," *2016 ACM/IEEE 7th International Conference on Cyber-Physical Systems (ICCPS)*, 2016, pp. 1–10.

7. S. Lin, B. Miller, J. Durand, G. Bleakley, A. Chigani, R. Martin, B. Murphy, and M. Crawford, "The industrial internet of things volume G1: reference architecture" *Industrial Internet Consortium*, pp. 10–46, 2017.
8. P. Rizwan, M. R. Babu, B. Balamurugan, and K. Suresh, "Real-time big data computing for Internet of Things and cyber physical system aided medical devices for better healthcare," *2018 Majan International Conference (MIC)*, pp. 1–8, 2018.
9. M. Lezzi, M. Lazoi, and A. Corallo, "Cybersecurity for industry 4.0 in the current literature: A reference framework," *Computers in Industry*, vol. 103, pp. 97–110, 2018.
10. W. Li, L. Xie, Z. Deng, and Z. Wang, "False sequential logic attack on scada system and its physical impact analysis," *Computers & Security*, vol. 58, pp. 149–159, 2016.
11. S. Karnouskos, "Stuxnet worm impact on industrial cyber-physical system security," *IECON 2011–37th Annual Conference of the IEEE Industrial Electronics Society*, pp. 4490–4494, IEEE, 2011.
12. W. Yu, T. Dillon, F. Mostafa, W. Rahayu, and Y. Liu, "Implementation of industrial cyber physical system: challenges and solutions," *2019 IEEE International Conference on Industrial Cyber Physical Systems (ICPS)*, pp. 173–178, 2019.
13. S. Lee, S. Lee, H. Yoo, S. Kwon, and T. Shon, "Design and implementation of cyber-security testbed for industrial IoT systems," *The Journal of Supercomputing*, vol. 74, no. 9, pp. 4506–4520, 2018.
14. F. Januario, C. Carvalho, A. Cardoso, and P. Gil, "Security challenges in scada systems over wireless sensor and actuator networks," *2016 8th International Congress on Ultra Modern Telecommunications and Control Systems and Workshops (ICUMT)*, pp. 363–368, IEEE, 2016.
15. Ray Y. Zhong, Chen Xu, Chao Chen, George Q. Huang, "Big data analytics for physical internet-based intelligent manufacturing shop floors," *International Journal of Production Research*, vol. 55, pp. 2610–2621, 2017.
16. Available online: https://www.anblicks.com/blog/how-big-data-analytics-in-manufacturing-strengthens-the-industry (accessed on 10 June 2022).
17. S. Ryu, Y.-J. Kang, and H. Lee, "A study on detection of anomaly behavior in automation industry," *2018 20th International Conference on Advanced Communication Technology (ICACT)*, pp. 377–380, 2018.
18. M. M. H. Onik, C. -S. Kim, and J. Yang, "Personal data privacy challenges of the fourth industrial revolution," *2019 21st International Conference on Advanced Communication Technology (ICACT)*, pp. 635–638, 2019.
19. Thuy Duong Oesterreich, Frank Teuteberg, "Understanding the implications of digitisation and automation in the context of Industry 4.0: A triangulation approach and elements of a research agenda for the construction industry," *Computers in Industry*, vol. 83, pp. 121–139, 2016.
20. J. Butt, "A strategic roadmap for the manufacturing industry to implement industry 4.0," *Designs*, vol. 4, no. 2, p. 11, May 2020.
21. B. Bajic, A. Rikalovic, N. Suzic, and V. Piuri, "Industry 4.0 implementation challenges and opportunities: A managerial perspective," *IEEE Systems Journal*, vol. 15, no. 1, pp. 546–559, March 2021.

Index

Printed in the United States
by Baker & Taylor Publisher Services